Theodore Low De Vinne, Joseph Moxon

Moxon's Mechanick Exercises

Or, the doctrine of handy-works applied to the art of printing; a literal reprint in

two volumes of the first edition published in the year 1683

Theodore Low De Vinne, Joseph Moxon

Moxon's Mechanick Exercises
Or, the doctrine of handy-works applied to the art of printing; a literal reprint in two volumes of the first edition published in the year 1683

ISBN/EAN: 9783337426866

Printed in Europe, USA, Canada, Australia, Japan

Cover: Foto ©berggeist007 / pixelio.de

More available books at **www.hansebooks.com**

the true Effigies of Laurenz Ians. Koster. Delineated from his Monumentall Stone Statue, Erected at Harlem.

MEMORIÆ
SACRVM

LAVRENTIO
COSTERO,
HARLEMENSI,
ALTERI CADMO,
ET ARTIS
TYPOGRAPHICÆ
CIRCA AN.DOM.
M.CCCC.XXX
INVENTORI
PRIMO,

BENE DE LITERIS
AC TOTO ORBE
MERENTI HANC
C.L.C.C.

STATVAM QVIA
ÆREA AVT MAR-
MOREA DE FVIT,
IRO MONVMEN-
TO POSVIT CIVIS
GRATISSIMVS

PETRVS
SCRIVERIVS
1635.

A

The true Effigies of Iohn Guttemberg Delineated from the Original Painting at Mentz in Germanie.

MOXON'S MECHANICK EXERCISES.

OR THE DOCTRINE OF HANDY-WORKS APPLIED TO THE ART OF

PRINTING

A LITERAL REPRINT IN TWO VOLUMES OF
THE FIRST EDITION PUBLISHED IN THE YEAR 1683

WITH PREFACE AND NOTES BY
THEO. L. DE VINNE

VOLUME I

NEW-YORK
THE TYPOTHETÆ OF THE CITY OF NEW-YORK
MDCCCLXXXXVI

PREFACE

JOSEPH MOXON was born at Wakefield in Yorkshire, England, August 8, 1627. There is no published record of his parentage or his early education. His first business was that of a maker and vender of mathematical instruments, in which industry he earned a memorable reputation between the years 1659 and 1683. He was not content with this work, for he had leanings to other branches of the mechanic arts, and especially toward the designing of letters and the making of printing-types.

In 1669 he published a sheet in folio under the heading of " Prooves of the Several Sorts of Letters Cast by Joseph Moxon." The imprint is " Westminster, Printed by Joseph Moxon, in Russell street, at the Sign of the Atlas, 1669." This specimen of types seems to have been printed, not to show his dexterity as a type-founder, but to advertise himself as a dealer in mathematical and scien-

B

tific instruments. The reading matter of the sheet
describes " Globes Celestial and Terrestrial, Large
Maps of the World, A Tutor to Astronomic and to
Geographie " — all of his own production. Reed
flouts the typography of this sheet: " It is a sorry
performance. Only one fount, the Pica, has any
pretensions to elegance or regularity. The others
are so clumsily cut or badly cast, and so wretch-
edly printed, as here and there to be almost unde-
cipherable." [1] The rude workmanship of these early
types proves, as he afterward admitted, that he had
never been properly taught the art of type-founding; that he had learned it, as he said others had,
"of his own genuine inclination."

It was then a difficult task to learn any valuable
trade. The Star Chamber decree of 1637 ordained
that there should be but four type-founders for the
kingdom of Great Britain, and the number of their
apprentices was restricted. When the Long Parlia-
ment met in 1640, the decrees of the Star Chamber
were practically dead letters, and for a few years
there was free trade in typography. In 1644 the
Star Chamber regulations were reimposed; in 1662
they were made more rigorous than ever. The im-
portation of types from abroad without the consent
of the Stationers' Company was prohibited. British

[1] "A History of the Old English Letter Foundries, with Notes
Historical and Biographical on the Rise and Progress of English
Typography." By Talbot Baines Reed, London, 1887, p. 181.

printers were compelled to buy the inferior types of
English founders, who, secure in their monopoly, did
but little for the improvement of printing.[1]

It is probable that the attention of Moxon was
first drawn to type-founding by the founders themselves, who had to employ mechanics of skill for
the making of their molds and other implements of
type-casting. In this manner he could have ob-
tained an insight into the mysteries of the art that
had been carefully concealed. He did not learn
type-making or printing in the usual routine. The
records of the Company of Stationers do not show
that he was ever made a freeman of that guild, yet
he openly carried on the two distinct businesses of
type-founding and printing after 1669. It is probable that he had a special permit from a higher
authority, for in 1665 he had been appointed hydrographer to the king, and a good salary was given
with the office. He was then devoted to the practical side of scientific pursuits, and was deferred to
as a man of ability.

He published several mathematical treatises between the years 1658 and 1687; one, called "Com-
pendium Euclidis Curiosi," was translated by him

---

[1] The four founders appointed by the Star Chamber did not
thrive. One of them, Arthur Nicholls, said of himself: "Of so
small benifitt hath his Art bine that for 4 yeares worke and practice he hath not taken above 48£, and had it not bine for other
imploymente he might have perrisht." Reed, p. 168.

from Dutch into English, and printed in London in 1677. Mores supposes that he had acquired a knowledge of Dutch by residence in Holland, but intimates that he was not proficient in its grammar.[1]

In 1676 he published a book on the shapes of letters, with this formidable title: " Regulæ Trium Ordinum Literarum Typographicarum; or the Rules of the Three Orders of Print Letters, viz: the Roman, Italick, English—Capitals and Small; showing how they are Compounded of Geometrick Figures, and mostly made by Rule and Compass. Useful for Writing Masters, Painters, Carvers, Masons and others that are Lovers of Curiosity. By Joseph Moxon, Hydrographer to the King's Most Excellent Majesty. Printed for Joseph Moxon on Ludgate Hill, at the Sign of Atlas, 1676." He then dedicated the book to Sir Christopher Wren, " as a lover of rule and proportion," or to one who might be pleased with this attempt to make alphabetical letters conform to geometric rules.

There is no intimation that the book was intended for punch-cutters. It contains specific directions about the shapes of letters, covering fifty-two pages, as proper introduction to the thirty-eight pages of model letters that follow, rudely drawn and printed from copper plates. Moxon says that these model

---

[1] " A Dissertation upon English Typographical Founders and Founderies." By Edward Rowe Mores, A. M. & A. S. S. [London], 1778. 8vo, p. 43.

letters are his copies of the letters of Christopher
Van Dijk, the famous punch-cutter of Holland. He
advises that each letter should be plotted upon a
framework of small squares—forty-two squares in
height and of a proportionate width, as is distinctly
shown in the plates of letters in this book.[1] Upon
these squares the draftsman should draw circles,
angles, and straight lines, as are fully set forth in
the instructions.

These diagrams, with their accompanying instruction, have afforded much amusement to type-founders.
All of them unite in saying that the forming of letters by geometrical rule is absurd and impracticable.
This proposition must be conceded without debate,
but the general disparagement of all the letters, in
which even Reed joins, may be safely controverted.
It is admitted that the characters are rudely drawn,
and many have faults of disproportion; but it must
not be forgotten that they were designed to meet
the most important requirement of a reader—to be
read, and read easily. Here are the broad hair-line,
the stubby serif on the lower-case and the brack-
eted serif on the capitals, the thick stem, the strong
and low crown on letters like m and n, with other
peculiarities now commended in old-style faces and
often erroneously regarded as the original devices
of the first Caslon. The black-letter has more merit

[1] See plates Nos. 11 to 17.

than the roman or italic. Some of the capitals are really uncouth; but with all their faults the general effect of a composition in these letters will be found more satisfactory to the bibliophile as a text-type than any form of pointed black that has been devised in this century as an improvement.

Moxon confesses no obligation to any one for his geometrical system, but earlier writers had propounded a similar theory. Books on the true proportions of letters had been written by Fra Luca Paccioli, Venice, 1509; Albert Dürer, Nuremberg, 1525; Geofroy Tory, Paris, 1529; and Yciar, Saragossa, 1548. Nor did the attempt to make letters conform to geometrical rules end with Moxon. In 1694, M. Jaugeon, chief of the commission appointed by the Academy of Sciences of Paris, formulated a system that required a plot of 2304 little squares for the accurate construction of every full-bodied capital letter. The manuscript and diagrams of the author were never put in print, but are still preserved in the papers of the Academy.

This essay on the forms of letters seems to have been sent out as the forerunner of a larger work on the theory and practice of mechanical arts. Under the general title of "Mechanick Exercises," in 1677, he began the publication, in fourteen monthly numbers, of treatises on the trades of the smith, the joiner, the carpenter, and the turner. These constitute the first volume of the "Mechanick Exercises."

The book did not find as many buyers as had been
expected. Moxon attributed its slow sale to political
excitement, for the Oates plot put the buying and
study of trade books away from the minds of read-
ers. He had to wait until 1683 before he began
the publication of the second volume, which con-
sists of twenty-four numbers, and treats of the art
of printing only. It is this second volume that is
here reprinted, for the first volume is of slight interest to the printer or man of letters.

Moxon's book has the distinction of being not
only the first, but the most complete of the few
early manuals of typography. Fournier's " Manuel
Typographique " of 1764 is the only book that can
be compared with it in minuteness of detail concerning type-making, but he treats of type-making
only. Reed says: " Any one acquainted with the
modern practice of punch-cutting cannot but be
struck, on reading the directions laid down in the
'Mechanick Exercises,' with the slightness of the
changes which the manual processes of that art
have undergone during the last two centuries. Indeed, allowing for improvements in tools, and the
greater variety of gauges, we might almost assert
that the punch-cutter of Moxon's day knew scarcely
less than the punch-cutter of our day, with the ac-
cumulated experience of two hundred years, could
teach him. . . . For almost a century it remained
the only authority on the subject; subsequently it

formed the basis of numerous other treatises both at home and abroad; and to this day it is quoted and referred to, not only by the antiquary, who desires to learn what the art once was, but by the practical printer, who may still on many subjects gather from it much advice and information as to what it should still be."[1]

During his business life, Moxon stood at the head of the trade in England. He was selected to cut a font of type for an edition of the New Testament in the Irish language, which font was afterward used for many other books. He cut also the characters designed by Bishop John Wilkins for his "Essay towards a Real Character and a Philosophical Language," and many mathematical and astronomical symbols. Rowe Mores, who describes him as an excellent artist and an admirable mechanic, says that he was elected a Fellow of the Royal Society in 1678.[2] There is no known record of the date of his death. Mores gives the year 1683 as the date of his relinquishment of the business of type-making, but he was active as a writer and a publisher for some years after.

The first volume of the "Mechanick Exercises," concerning carpentry, etc., went to its third edition in 1703, but the second volume, about printing, has been neglected for two centuries. During this long

[1] Reed, "Old English Letter Foundries," pp. 185, 186.
[2] Mores, "English Founders," p. 42.

interval many copies of the first small edition of five hundred copies have been destroyed. A perfect copy is rare, and commands a high price, for no early book on technical printing is in greater request.[1]

The instruction directly given is of value, but bits of information indirectly furnished are of greater interest. From no other book can one glean so many evidences of the poverty of the old printing-house. Its scant supply of types, its shackly hand-presses, its mean printing-inks, its paper windows and awkward methods, when not specifically confessed, are plainly indicated. The high standard of proof-reading here exacted may be profitably contrasted with its sorry performance upon the following pages. The garments worn by the workmen are shown in the illustrations. Some of the quainter usages of the trade are told in the "Customs of the Chappel," and those of the masters, in the ceremonies of the Stationers' Company, and in the festivals in which masters and workmen joined. To the student of printing a reading of the book is really necessary for a clear understanding of the mechanical side of the art as practised in the seventeenth century.

[1] Hansard says ("Typographia," p. vii) : " I have never been able to meet with more than two copies of this work — one in the Library of the British Museum—the other in the Library of the Society of Arts." The writer knows of but three copies in America: one in the Library Company of Philadelphia; one in the Library of the Typothetæ of New-York ; one in his own collection.

c

NOTE BY THE PRINTER

This edition of the "Mechanick Exercises" is a line-for-line and page-for-page reprint of the original text. The only suppression is that of the repetition of the words "Volume II" in the running title and the sub-titles, which would unnecessarily mislead the reader, and of the old signature marks that would confuse the bookbinder. Typographic peculiarities have been followed, even to the copying of gross faults, like doublets, that will be readily corrected by the reader. The object of the reprint is not merely to present the thought of the author, but to illustrate the typographic style of his time with its usual defects. A few deviations from copy that seemed to be needed for a clearer understanding of the meaning of the author have been specified at the end of the second volume. The irregular spelling and punctuation of the copy, its capricious use of capitals and italic, its headings of different sizes of type, have been repeated. At this point imitation has stopped. Turned and broken letters, wrong-font characters, broken space-lines, and bent rules have not been servilely reproduced. These blemishes, as well as the frequent "monks" and "friars" in the presswork, were serious enough to prevent an attempt at a photographic facsimile of the pages.

The two copies of Moxon that have served as "copy" for this reprint show occasional differences in spelling and punctuation. Changes, possibly made in the correction of batters, or after the tardy discovery of faults, must have been done while the form was on press and partly printed. The position of the plates differs seriously in the two copies; they do not follow each other in the numerical order specified. In this reprint the plates that describe types and tools have been placed near their verbal descriptions.

The type selected for this work was cast from matrices struck with the punches (made about 1740) of the first Caslon. It is of the same large English body as that of the original, but a trifle smaller as to face, and not as compressed as the type used by Moxon; but it repeats many of his peculiarities, and fairly reproduces the more important mannerisms of the printing of the seventeenth century.

The portraits have been reproduced by the artotype process of Bierstadt; the descriptive illustrations are from the etched plates of the Hagopian Photo-Engraving Company.

DUCTOR ad
ASTRONOMIAM
&
GEOGRAPHIAM
Vel USUS
GLOBI
Cœlestis quam Terrestris
in Libris sex.
Astron: & Geog: Rud:
Astron & Geogr:
Nautica.
Vt Astrologica.
Gnomonica.
Spheric Triang:
per Josephum Moxon

Joseph Moxon.
Born at Wakefeild *August* 9.
Anno 1627.

MECHANICK EXERCISES:

Or, the Doctrine of

Handy-works.

Applied to the Art of

Printing.

By *Joseph Moxon*, Member of the Royal Society, and *Hydrographer* to the King's Most Excellent Majesty.

LONDON.

Printed for *Joseph Moxon* on the West-side of *Fleet-ditch*, at the Sign of *Atlas*. 1 6 8 3.

To the Right Reverend Father in GOD, *JOHN* Lord Bifhop of *Oxford*, and Dean of *Chrift-Church*; And to the Right Honourable Sir *LEOLINE JENKINS* Knight, and Principal Secretary of State; And to the Right Honourable Sir *JOSEPH WILLIAM-SON* Knight; and one of His Majefties moft Honourable Privy-Council.

Right Honourable.

*Y*our ardent *affections to promote* Typographie *has eminently appeared in the great Charge you have been at to make it famous* here in England; *whereby this Royal Ifland ftands particularly obliged to your Generous and Publick Spirits, and the whole Common-Wealth of Book-men throughout the World, to your Candid Zeal for the promulgation of good Learning.*

Wherefore I humbly Dedicate this Piece of Typographie *to your Honours; and*

as it is (I think) the first of this na-
ture, fo I hope you will favourably excufe
fmall Faults in this Undertaking ; for great
ones I hope there are none, unlefs it be in this
prefumptuous Dedication; *for which I hum-*
bly beg your Honours pardon : Subfcribing
my self, My Lord and Gentlemen,

Your Honours moft Humble
and Obedient Servant.

Jofeph Moxon.

MECHANICK

MECHANICK EXERCISES:

Or, the Doctrine of

𝕳andy=works.

Applied to the Art of

𝕻rinting.

P R E F A C E.

BEfore I begin with Typographie, *I shall say some-what of its Original Invention; I mean here in* Europe, *not of theirs in* China *and other Eastern Countries, who (by general assent) have had it for many hundreds of years, though their Invention is very different from ours; they Cutting their Letters upon Blocks in whole Pages or Forms, as among us our Wooden Pictures are Cut; But* Printing *with single Letters Cast in Mettal, as with us here in* Europe, *is an Invention scarce above Two hundred and fifteen years old; and yet an undecidable Controversie about the original Contriver or Contrivers remains on foot,*

between

between the Harlemers *of* Holland, *and thofe of* Mentz
in Germany : *But becaufe the difference cannot be deter-
min'd for want of undeniable Authority, I fhall only deli-
ver both their Pleas to this* Scientifick Invention.

The Harlemers *plead that* Lawrenfz Janfz Kofter of
Harlem *was the firft Inventer of* Printing, *in the year
of our Lord* 1430. *but that in the Infancy of this Invention he ufed only Wooden Blocks (as in* China, *&c. aforefaid)
but after fome time he left off Wood, and Cut fingle Letters
in Steel, which he funck into Copper* Matrices, *and fitting
them to Iron Molds, Caft fingle Letters of Mettal in thofe*
Matrices.   *They fay alfo, that his Companion,* John
Gutenberg, *ftole his Tools away while he was at Church,
and with them went to* Mentz *in* Germany, *and there
fet his Tools to work, and promoted His claim to the firft
Invention of this Art, before* Kofter *did His.*

To prove this, they fay that Rabbi Jofeph (*a Jew*)
in his Chronicle, mentions a Printed Book that he faw in
Venice, *in the year* 5188. *according to the Jewifh Account, and by ours the year* 1428. *as may be read in*
Pet. Scriverius.

They fay much of a Book intituled De Spiegel, *Printed
at* Harlem *in* Dutch *and* Latin ; *which Book is yet there
to be feen* : *and they alledge that Book the firft that ever
was Printed* : *But yet fay not when this Book was Printed.*

*Notwithftanding this Plea, I do not find (perhaps be-
caufe of their imperfect Proofs) but that* Gutenberg *of*
Mentz *is more generally accepted for the firft Inventer of*
Printing, *than* Kofter *of* Harlem.

The Learned Dr. Wallis *of* Oxford, *hath made an
Inquiry into the original of this Invention, and hath in
brief fum'd up the matter in thefe words.*

About

About the year of our Lord 1460. The Art of *Printing* began to be invented and practifed in *Germany*, whether firft at *Mentz* or firft at *Harlem* it is not agreed: But it feems that thofe who had it in confideration before it was brought to perfection, difagreeing among themfelves, did part Company; and fome of them at *Harlem*, others at *Mentz* perfued the defign at the fame time.

The Book which is commonly reputed to have been firft Printed is, Tullies Offices, *of which there be Copies extant* (*as a Rarity*) *in many Libraries; which in the clofe of it is faid to be Printed at* Mentz, *in the year of our Lord* 1465. (*fo fays that Copy in the* Bodleyan *Library*) *or* 1466. (*fo that in the Library of* Corpus Chrifti.) *The words in the clofe of that in* Corpus Chrifti *Colledge* Oxon *are thefe,*

Præfens *Marcij Tullij* Clariffimum opus, *Johanes Huft, Moguntinus* Civis, non Atrimento, plumali canna, neq; ærea, fed Arte quadam perpulchra, *Petri* manu *Petri* de *Geurfhem* pueri mei, feliciter effeci, finitum Anno M CCCC LX VI quarto die Menfis *Februarij.*

The like in the Bodleyan *Library; fave there the Date is only thus,* Finitum Anno M CCCC LX V. *In the fame Book there are thefe written Notes fubjoyned:* Hic eft ille *Johannes Fauftus,* coadjutor *Johannes Gutenbergij* primi *Typographiæ* inventaris, Alter coadjuto erat *Petrus Schæfer,* i. Opilio. Quovix.

Cælando promptior alter erat, inquit *Johan. Arnoldus* in Libello de Chalcographiæ inventione, *Scheffer* primas finxit quas vocant *Matrices.* Hi tres exercuerunt artem primo in communi. mox rupto fœdere feorfim fibi quifq; privatim.

And

And again (*in a later hand*) Inventionem artis *Typographicæ* ad Annum 1453. aut exerciter referunt Sabillicus *En.* 10.lib.6.&*Monfterus.* Alij ad Annum 1460. Vide *Polid. Virg.*lib. 2. de Invent. Rerum, *Theod. Bibland. de Ratione communis linguarum.* cap. de *Chalcographia.*

At Harlem *and fome other places in* Holland, *they pretend to have Books Printed fomewhat ancienter than this*; *but they are moft of them* (*if not all*) *done by way of Carving whole Pages in Wood, not by fingle Letters Caft in Mettal, to be Compofed and Diftributed as occafion ferves, as is now the manner.*

The chief Inventer at Harlem *is faid to be* Laurens Janfz Kofter.

After thefe two places (Mentz *and* Harlem) *it feems next of all to have been practifed at* Oxford: *For by the care, and at the charge of King* Henry *the 6th, and of* Thomas Bourchier *then Arch-Bifhop of* Canterbury (*and Chancellour of the Univerfity of* Oxford) Robert Turner *Mafter of the Robe, and* William Caxton *a Merchant of* London *were for that purpofe fent to* Harlem, *at the charges partly of the King, partly of the Arch-Bifhop, who then* (*becaufe thefe of* Harlem *were very chary of this fecret*) *prevailed privately with* one Frederick Corfeles *an under-Workman, for a fum of Money, to come over hither*; *who thereupon did at* Oxford *fet up the Art of* Printing, *before it was exercifed any where elfe in* England, *or in* France, Italy, Venice, Germany, *or any other place, except only* Mentz *and* Harlem (*aforementioned*): *And there be feveral Copies yet extant* (*as one in the Archives of the Univerfity of* Oxford, *another in the Library of Dr.* Tho. Barlow, *now Bifhop of* Lincoln) *of a Treatife*
of

of St. Jerome *(as it is there called (becaufe found a-
mong St.* Jerom's *Works) or rather* Ruffinus *upon the
Creed, in a broad* Octavo) *Printed at* Oxford *in the
year* 1468. *as appears by the words in the clofe of it.*

Explicit expofitio Sancti *Jeronimi* in fembolo Apo-
ftolorum ad papam Laurentium Impreffi *Oxonie* &
finita Anno Domini M CCCC LX VIII. xvij die *De-
cembris.*

Which is but three years later than that of Tullies
Offices *at* Mentz, *in* 1465. *and was perhaps one of the
firft Books Printed on Paper; (that of* Tully *being on
Vellom.) And there the excercife of* Printing *hath
continued fucceffively to this day.*

Soon after William Caxton *(the fame I fuppofe who
firft brought it to* Oxford) *promoted it to* London *alfo, which* Baker *in his Chronicle (and fome others) fay
to have been about the year* 1471. *but we have fcarce a-
ny Copies of Books there Printed remaining (that I have
feen) earlier than the year* 1480. *And by that time, or
foon after, it began to be received in* Venice, Italy,
Germany, *and other places, as appears by Books yet extant, Printed at divers places in thofe Times. Thus far
Dr.* Wallis.

*But whoever were the Inventers of this Art, or (as
fome Authors will have it)* Science; *nay, Science of Sciences (fay they) certain it is, that in all its Branches it
can be deemed little lefs than a Science: And I hope I fay
not to much of* Typographie: *For Dr.* Dee, *in his
Mathematical Preface to* Euclids Elements of Geome-
trie, *hath worthily taken pains to make* Architecture *a
Mathematical Science; and as a vertual Proof of his own
Learned Plea, quotes two Authentique Authors, viz.*
Vitruvius

1*

Vitruvius *and* Leo Baptifta, *who both give their de-*
fcriptions and applaufe of Architecture: *His Arguments*
are fomewhat copious, and the Original eafily procurable
in the Englifh Tongue; therefore inftead of tranfcribing
it, I fhall refer my Reader to the Text it felf.

Upon the confideration of what he has faid in behalf
of Architecture, *I find that a* Typographer *ought*
to be equally quallified with all the Sciences that be-
comes an Architect, *and then I think no doubt re-*
mains that Typographie *is not alfo a Mathematical*
Science.

For my own part, I weighed it well in my thoughts, and
find all the accomplifhments, and fome more of an Architect *neceffary in a* Typographer: *and though my bufinefs*
be not Argumentation, yet my Reader, by perufing the
following difcourfe, may perhaps fatisfie himfelf, that a
Typographer *ought to be a man of Sciences.*

By a Typographer, *I do not mean a* Printer, *as he is*
Vulgarly accounted, any more than Dr. Dee *means a* Carpenter *or* Mafon *to be an* Architect: *But by a* Typographer, *I mean fuch a one, who by his own Judgement,*
from folid reafoning with himfelf, can either perform, or
direct others to perform from the beginning to the end,
all the Handy-works and Phyfical Operations relating to
Typographie.

Such a Scientifick *man was doubtlefs he who was*
the firft Inventer of Typographie; *but I think few*
have fucceeded him in Science, though the number of
Founders *and* Printers *be grown very many: Infomuch*
that for the more eafie managing of Typographie, *the*
Operators have found it neceffary to devide it into feveral
Trades, each of which (in the ftricteft fence) ftand no
nearer

nearer related to Typographie, *than* Carpentry *or* Ma-
ſonry, *&c. are to* Architecture. *The ſeveral deviſions
that are made, are,*

Firſt *The* Maſter Printer, *who is as the Soul of* Printing ; *and all the Work-men as members of the Body govern-
ed by that Soul ſubſerveient to him* ; *for the* Letter-Cutter
would Cut no Letters, the Founder *not ſinck the* Matrices,
or Caſt and Dreſs the Letters, the Smith *and* Joyner *not
make the* Preſs *and other Utenſils for* Printing, *the* Compoſiter *not Compoſe the Letters, the* Correcter *not read
Proves, the* Preſs-man *not work the Forms off at the*
Preſs, *or the* Inck-maker *make* Inck *to work them
with, but by Orders from the* Maſter-Printer.

Secondly, The Letter-Cutter, ⎫
Thirdly, The Letter-Caſter, ⎬ Founders.
Fourthly, The Letter-Dreſſer. ⎭

But very few Founders *exerciſe, or indeed can perform
all theſe ſeveral Trades ; though each of theſe are indifferently called* Letter-Founders.

Fifthly, The Compoſiter, ⎫
Sixthly, The Correcter, ⎬ Printers.
Seventhly, The Preſs-man, ⎭
Eighthly, The Inck-maker.

*Beſides ſeveral other Trades they take in to their Aſ-
ſiſtance* ; *as the* Smith, *the* Joyner, *&c.*

ADVER-

ADVERTISEMENT.

THE continuation of my setting forth *Mechanick Exercises* having been obstructed by the breaking out of the Plot, which took off the minds of my few Customers from buying them, as formerly; And being of late much importun'd by many worthy Persons to continue them; I have promised to go on again, upon Condition, That a competent number of them may be taken off my hand by Subscribers, soon after the publication of them in the *Gazet*, or posting up Titles, or by the *Mercurius Librarius*, &c.

Therefore such Gentlemen or others as are willing to promote the coming forth of these *Exercises*, are desired to Subscribe their Names and place of abode: That so such Persons as live about this City may have them sent so soon as they come forth: Quick Sale being the best encouragement.

Some Gentlemen (to whom they are very acceptable) tell me they will take them when all *Trades* are finish't, which cannot reasonably be expected from me (my Years considered) in my life-time; which implies they will be Customers when I'me dead, or perhaps by that time some of themselves.

The price of these Books will be 2*d.* for each Printed Sheet. And 2*d.* for every Print taken off of Copper Cuts.

There are three reasons why this price cannot be thought dear.

1. The Writing is all new matter, not Collected, or Translated from any other Authors: and the drafts of the Cuts all drawn from the Tools and Machines used in each respective Trade.

2. I Print but 500 on each Sheet, And those upon good Paper: which makes the charge of Printing dear, proportionable to great numbers.

3. Some Trades are particularly affected by some Customers, (who desire not the rest,) and consequently sooner sold off, which renders the remainder of the un-sold *Exercises* unperfect, and therefore not acceptable to such as desire all: so that they will remain as waste-Paper on my hands.

JOSEPH MOXON.

MECHANICK EXERCISES:

Or, the Doctrine of

ℌandy=works.

Applied to the Art of

ℌrinting.

§ 2. *Of the Office of a Master*-Printer.

I Shall begin with the Office of a *Master-Printer*, becaufe (as aforefaid) he is the Director of all the Work men, he is the Bafe (as the *Dutchmen* properly call him) on which the Workmen ftand, both for providing Materials to Work withal, and fucceffive variety of Directions how and in what manner and order to perform that Work.

His Office is therefore to. provide a Houfe, or Room or Rooms in which he is to fet his *Printing-Houfe*. This expreffion may feem ftrange, but it is *Printers* Language: For a *Printing-Houfe* may admit of a twofold meaning; one the Vulgar acceptance, and

and is relative to the Houſe or Place wherein *Printing* is uſed; the other a more peculiar Phraſe *Printers* uſe among themſelves, *viz.* only the *Printing* Tools, which they frequently call a *Printing-Houſe*: Thus they ſay, Such a One has ſet up a *Printing-Houſe*, when as thereby they mean he has furniſh'd a Houſe with *Printing* Tools. Or ſuch a one has remov'd his *Printing-Houſe*, when thereby they only mean he has remov'd the Tools us'd in his former Houſe. Theſe expreſſions have been uſed Time out of mind, and are continued by them to this day.

But to proceed, Having conſider'd what number of *Preſſes* and *Caſes* he ſhall uſe, he makes it his buſineſs to furniſh himſelf with a Room or Rooms well-lighted, and of convenient capacity for his number of *Preſſes* and *Caſes*, allowing for each *Preſs* about Seven Foot ſquare upon the Floor, and for every *Frame* of *Caſes* which holds Two pair of *Caſes, viz.* one pair *Romain* and one pair *Itallica*, Five Foot and an half in length (for ſo much they contain) and Four Foot and an half in breadth, though they contain but Two Foot and Nine Inches: But then room will be left to paſs freely between two *Frames.*

We will ſuppoſe he reſolves to have his *Preſſes* and *Caſes* ſtand in the ſame Room (though in *England* it is not very cuſtomary) He places the *Caſes* on that ſide the Room where they will moſt conveniently ſtand, ſo, as when the *Compoſiter* is at work the Light may come in on his Left-hand; for elſe his Right-hand plying between the Window-light and his Eye might ſhadow the *Letter* he would pick up: And the *Preſſes* he places ſo, as the Light may fall from a Window
right

right before the *Form* and *Tinpan*: And if fcituation will allow it, on the North-fide the Room, that the *Prefs-men*, when at their hard labour in *Summer* time, may be the lefs uncommoded with the heat of the *Sun*: And alfo that they may the better fee by the conftancy of that Light, to keep the whole *Heap* of an equal Colour.

He is alfo to take care that his *Preffes* have a folid and firm Foundation, and an even Horizontal Floor to ftand on, That when the *Preffes* are fet up their Feet fhall need no Underlays, which both damage a *Prefs*, are often apt to work out, and confequently fubject it to an unftable and loofe pofition, as fhall further be fhewn when we come to the Setting up of the *Prefs*.

And as the Foundation ought to be very firm, fo ought alfo the Roof and Sides of the *Prefs Room* to be, that the *Prefs* may be faftned with Braces over-head and on its Sides, as well and fteddy as under foot.

He is alfo to take care that the Room have a clear, free and pretty lofty Light, not impeded with the fhadow of other Houfes, or with Trees; nor fo low that the Sky-light will not reach into every part of the Room: But yet not too high, left the violence of *Winter* (*Printers* ufing generally but Paper-windows) gain too great advantage of Freefing the Paper and Letter, and fo both Work and Workman ftand ftill. Therefore he ought to Philofophize with himfelf, for the making the height of his Lights to bear a ra-tional proportion to the capacity of the Room.

Here being but two fides of the Room yet ufed,
he

he places the *Correcting ftone* againft a good Light, and as near as he can towards the middle of the Room, that the *Compofiters* belonging to each end of the Room may enjoy an equal accefs to it. But fometimes there are feveral *Correcting-ftones* plac'd in feveral parts of the Room.

The *Lye-Trough* and *Rincing-Trough* he places towards fome corner of the Room, yet fo as they may have a good Light; and under thefe he caufes a *Sink* to be made to convey the Water out of the Room: But if he have other conveniencies for the placing thefe Troughs, he will rather fet them out of the Room to avoid the flabbering they caufe in.

About the middle of the Room he places the *Deftributing-Frame* (*viz.* the *Frame* on which the *Forms* are fet that are to be *Deftributed*) which may ftand light enough, though it ftand at fome confiderable diftance from the Window.

In fome other empty place of the Room (leaft frequented) he caufes fo many *Neft-Frames* to be made as he thinks convenient to hold the *Cafes* that may lye out of prefent ufe; and the *Letter-boards* with *Forms* fet by on them, that both the *Cafes* and the *Forms* may be the better fecured from running to *Pye*.

Having thus contrived the feveral Offices of the Room, He furnifhes it with *Letters*, *Preffes*, *Cafes*, *Chafes*, *Furniture*, &c. Of each of which in Order.

¶. 2. *Of*

¶. 2. *Of* Letter.

He provides a *Fount* (properly a *Fund*) of *Letter* of all Bodies; for moſt *Printing-Houſes* have all except the two firſt, *viz. Pearl, Nomparel, Brevier, Long-Primmer, Pica, Engliſh, Great-Primmer, Double-Pica, Two-Lin'd-Engliſh, Great-Cannon.*

Theſe are the *Bodies* moſt of uſe in *England*; But the *Dutch* have ſeveral other *Bodies*: which becauſe there is little and almoſt no perceivable difference from ſome of theſe mentioned, I think they are not worth naming. Yet we have one *Body* more which is ſometimes uſed in *England*; that is a *Small Pica*, but I account it no great diſcretion in a *Maſter-Printer* to provide it; becauſe it differs ſo little from the *Pica*, that unleſs the Workmen be carefuller than they ſometimes are, it may be mingled with the *Pica*, and ſo the Beauty of both *Founts* may be ſpoil'd.

Theſe aforeſaid *Bodies* are commonly *Caſt* with a *Romain, Italica*, and ſometimes an *Engliſh Face*. He alſo provides ſome *Bodies* with the *Muſick*, the *Greek*, the *Hebrew*, and the *Syriack Face*: But theſe, or ſome of theſe, as he reckons his oppertunities may be to uſe them.

And that the Reader may the better underſtand the ſizes of theſe ſeveral *Bodies*, I ſhall give him this Table following; wherein is ſet down the number of each *Body* that is contained in one Foot.

Pearl,

Pearl,	184	
Nomparel,	150	
Brevier,	112	
Long-Primmer,	92	
Pica,	75	contained in one Foot.
Englifh,	66	
Great-Primmer,	50	
Double-Pica,	38	
Two Lin'd Englifh,	33	
Great-Cannon.	17½	

His care in the choice of thefe *Letters* are,

Firft, That the *Letter* have a true fhape: Which he may know, as by the §. of *Letter-Cutting.*

I confefs this piece of Judgement, *viz.* knowing of true Shape, may admit of fome controverfy, becaufe neither the Ancients whom we received the knowledge of thefe *Letters* from, nor any other authentick Authority have delivered us Rules, either to make or know true fhape by: And therefore it may be objected that every one that makes *Letters* but tolerably like *Romain, Italick, &c.* may pretend his to be true fhap'd.

To this I anfwer, that though we can plead no Ancient Authority for the fhape of *Letters,* yet doubtlefs (if we judge rationally) we muft conclude that the *Romain Letters* were Originally invented and contrived to be made and confift of Circles, Arches of Circles, and ftraight Lines; and therefore thofe *Letters* that have thefe Figures, either entire, or elfe properly mixt, fo as the Courfe and Progrefs of the

Pen

Pen may beſt admit, may deſerve the name of true
Shape, rather than thoſe that have not.

Beſides, Since the late made *Dutch-Letters* are ſo generally, and indeed moſt deſervedly accounted the
beſt, as for their Shape, conſiſting ſo exactly of Mathematical Regular Figures as aforeſaid, And for the
commodious Fatneſs they have beyond other *Letters*, which eaſing the Eyes in Reading, renders them
more Legible; As alſo the true placing their Fats
and their Leans, with the ſweet driving them into
one another, and indeed all the accompliſhments
that can render *Letter* regular and beautiful, do
more viſibly appear in them than in any *Letters*
Cut by any other People: And therefore I think
we may account the Rules they were made by, to
be the Rules of true ſhap'd *Letters*.

For my own part, I liked their *Letters* ſo well, e-
ſpecially thoſe that were Cut by *Chriſtophel Van Dijck*
of *Amſterdam*, that I ſet my ſelf to examine the
Proportions of all and every the parts and Members
of every *Letter*, and was ſo well pleaſed with the
Harmony and Decorum of their Symetrie, and
found ſo much Regularity in every part, and ſo good
reaſon for his Order and Method, that I examined
the biggeſt of his *Letters* with Glaſſes, which ſo
magnified the whole *Letter*, that I could eaſily di-
ſtinguiſh, and with ſmall Deviders meaſure off the
ſize, ſcituation and form of every part, and the proportion every part bore to the whole; and for my
own future ſatisfaction collected my Obſervations into a Book, which I have inſerted in my *Exerciſes*
on *Letter-Cutting*. For therein I have exhibited to
the

the World the true Shape of *Chriſtophel Van Dijcks* aforeſaid *Letters,* largely Engraven in Copper Plates.

Whence I conclude, That ſince common conſent of Book-men aſſign the Garland to the *Dutch-Letters* as of late *Cut,* and that now thoſe *Letters* are reduced unto a Rule, I think the Objection is Anſwered; And our *Maſter-Printers* care in the choice of good and true ſhap'd *Letters* is no difficult Task: For if it be a large Bodied *Letter,* as *Engliſh, Great-Primmer* and upwards, it will ſhew it ſelf; and if it be ſmall, as *Pearl, Nomparel,* &c. though it may be difficult to judge the exact Symetry with the naked Eye, yet by the help of a *Magnifying-Glaſs* or two if occaſion be, even thoſe ſmall *Letters* will appear as large as the biggeſt Bodied *Letters* ſhall to the naked Eye: And then it will be no difficult Task to judge of the Order and Decorum even of the ſmalleſt Bodied *Letters.* For indeed, to my wonder and aſtoniſhment, I have obſerv'd *V. Dijcks Pearl Dutch Letters* in Glaſſes that have Magnified them to great *Letters,* and found the whole Shape bear ſuch true proportion to his great *Letters,* both for the *Thickneſs, Shape, Fats* and *Leans,* as if with Compaſſes he could have meaſur'd and ſet off in that ſmall compaſs every particular Member, and the true breadth of every *Fat* and *Lean Stroak* in each *Letter,* not to exceed or want (when magnified) of *Letter Cut* to the *Body* it was Magnified to.

His ſecond care in the choice of *Letters* is, That they be deep *Cut;* for then they will *Print* clear the longer, and be leſs ſubject to entertain *Picks.*

His third care, That they be deep ſunck in the
Matrices,

Matrices leaft the bottom line of a *Page* Beard. Yet though they be deep funk, His care ought to be to fee the Beard alfo well cut off by the *Founder*.

And a Fourth Care in the choice of *Letter* is, That his *Letter* be Caft upon good Mettal, that it may laft the longer.

Of each Body he provides a *Fount* fuitable to fuch forts of Work as he defigns to do; But he provides not an equal weight of every *Fount*; Becaufe all thefe Bodies are not in equal ufe: For the *Long-Primmer*, *Pica* and *Englifh* are the Bodies that are generally moft ufed; And therefore he provides very large *Founts* of thefe, *viz.* of the *Long-Primmer* in a fmall *Printing-Houfe*, Five hundred Pounds weight *Romain* and *Italica*, whereof One hundred and fifty Pounds may be *Italica*. Of the *Pica* and *Englifh*, *Roman* and *Italica*, Eight, Nine hundred, or a Thoufand Pounds weight: when as of other *Founts* Three or Four hundred Pounds weight is accounted a good *Fount*: And of the *Cannon* and *Great-Cannon*, One hundred Pounds or fomewhat lefs may ferve his turn; Becaufe the common ufe of them is to fet Titles with.

Befides *Letters* he Provides Characters of Aftronomical Signs, *Planets*, *Afpects*, *Algebraical* Characters, Phyfical and Chimical Characters, &c. And thefe of feveral of the moft ufed Bodies.

He Provides alfo *Flowers* to fet over the Head of a *Page* at the beginning of a Book: But they are now accounted old-fafhion, and therefore much out of ufe. Yet *Wooden-Borders*, if well Drawn, and neatly Cut, may be *Printed* in a Creditable Book, As alfo, *Wooden-*

2

Wooden-Letters well Drawn and neatly Cut may be used at the beginning of a *Dedication, Preface, Section, &c.* Yet instead of *Wooden Letters, Capitals* Cast in Mettal generally now serves; because but few or good *Cutters* in *Wood* appear.

He also provides *Brafs Rules* of about Sixteen Inches long, that the *Compositer* may cut them into such Lengths as his Work requires.

In the choice of his *Brafs Rules*, he examines that they be exactly *Letter* high; for if they be much too high, they may cut through *Paper, Tinpan* and *Blankets* too; And if they be but a little too high, not only the Sholder, or Beard, on either side them will *Print* black; but they will bear the *Plattin* off the *Letters* that stand near them, so that those *Letters* will not *Print* at all: And if they be too low, then the *Rules* themselves will not *Print*.

It sometimes happens through the unskilfulnefs of the *Joyner*, (for they commonly, but unproperly, imploy *Joyners* to make them) that a Length shall be hollow in the middle both on the Face and Foot, and shall run driving higher and higher towards both ends: Hence it comes to pafs, that when the *Compositer* cuts a piece of *Rule* to his intended Length, the *Rule* shall *Print* hard at one end, and the other shall not *Print* at all; So that he shall be forced to knock up the foot of the low end, as shall be shewn in its proper place.

But the careful *Master-Printer* having found that his *Brafs Rules* is *Letter* high all the whole Length, will also examine whether it be straight all the whole Length, which he does by applying both the
Face

Plate 1

A

B

A B C D E F G H I K L M N O P Q R S T U V W X Y Z 1 2 3 4 5 6 7 8 9 0

a b c d e f g h i j k l m n o p q r s t u v w x y z

Face and Foot to the furface of the *Correcting-ftone*;
And if the Face and Foot comply fo clofely with the
Correcting-ftone, that light cannot be feen between
them, he concludes the *Brafs-Rule* is ftraight.

Then he examines the Face or Edge of the *Rule*,
whether it have an Edge of an equal breadth all the
whole Length, and that the Edge be neither too thick
nor too fine for his porpofe.

He fhould alfo take care that the *Brafs*, before it
be cut out, be well and skilfully Planifh't, nor would
that charge be ill beftowd; for it would be faved
out of the thicknefs of the *Brafs* that is commonly
ufed: For the *Joyners* being unskilful in Planifhing,
buy Neal'd thick *Brafs* that the *Rule* may be ftrong
enough, and fo cut it into flips without Hammering,
which makes the *Rule* eafily bow any way and ftand
fo, and will never come to fo good and fmooth an
Edge as Planifh't *Brafs* will. Befides, *Brafs* well Planifh't will be ftiffer and ftronger at half the thicknefs
than unplanifh't *Brafs* will at the whole: As I fhall
further fhew when I come to Exercife upon *Mathematical Inftrument-making*.

§. 3. *Of* Cafes.

Next he provides *Cafes*. A Pair of *Cafes* is an *Upper-
Cafe* and a *Lower Cafe*.

The *Upper Cafe* and the *Lower-Cafe* are of an equal
length, breadth and depth, *viz.* Two Foot nine Inches
long, One Foot four Inches and an half broad, and
about an Inch and a quarter deep, befides the bottom
Board; But for fmall Bodied *Letters* they are made
fomewhat fhallower, and for great Bodies deeper.

Long-

Long-Primmer and downwards are accounted fmall Bodies; *Englifh* and upwards are accounted great Bodies.

The conveniencies of a fhallow *Cafe* is, that the *Letters* in each Box lye more vifible to the laft, as being lefs fhadowed by the fides of the *Boxes*.

The conveniencies of a deep *Cafe* is, that it will hold a great many *Letters*, fo that a *Compofiter* needs not fo often *Deftribute*. 2dly. It is not fo foon *Low*, (as *Compofiters* fay when the *Cafe* grows towards empty) and a *Low Cafe* is unconvenient for a *Compofiter* to work at, partly becaufe the *Cafe* ftanding fhelving downwards towards them, the *Letters* that are in the *Cafe* tend towards the hither fide of the *Cafe*, and are fhadowed by the hither fide of that *Box* they lye in, fo that they are not fo eafily feen by the Eye, or fo ready to come at with the Fingers, as if they lay in the middle of the *Box*.

Thefe *Cafes* are encompaffed about with a *Frame* about Three quarters of an Inch broad, that the ends of the feveral partitions may be let into the fubftance of the *Frame*: But the hithermoft fide of the *Frame* is about half an Inch higher than the other fides, that when either the *Galley* or another pair of *Cafes* are fet upon them, the bottom edge of the *Galley*, or of thofe *Cafes* may ftop againft that higher *Frame*, and not flide off.

Both the *Upper* and the *Lower Cafe* have a thick Partition about three quarters of an Inch broad, Duff-tail'd into the middle of the upper and under Rail of the *Frame*. This Partition is made thus broad, that Grooves may be made on either fide of it to receive

ceive the ends of thofe Partitions that devide the breadth of the *Cafe*, and alfo to ftrengthen the whole *Frame*; for the bottom Board is as well nailed to this thick Partition as to the outer *Frame* of the *Cafe*.

But the devifions for the feveral *Boxes* of the *Upper* and *Lower Cafes* are not alike: for each half of the whole length of the *Upper-Cafe* is devided into feven equal parts, as you may fee in Plate 1. at A, and its breadth into feven equal parts, fo that the whole *Upper-Cafe* is divided into Ninety eight fquare *Boxes*, whofe fides are all equal to one another.

But the Two halfs of the length of the *Lower-Cafe* are not thus devided; for each half of the length of the *Lower-Cafe* is devided into Eight equal parts, and its breadth into Seven; but it is not throughout thus devided neither; for then the *Boxes* would be all of equal fize: But the *Lower-Cafe* is devided into four feveral fizes of *Boxes*, as you may fee in Plate 1. B.

The reafon of thefe different fizes of *Boxes* is, That the biggeft *Boxes* may be difpofed neareft the *Compofiters* hand, becaufe the Englifh Language, and confequently all Englifh *Coppy* runs moft upon fuch and fuch Sorts; fo that the *Boxes* that holds thofe Sorts ought to be moft capacious.

His care in the choice of thefe *Cafes* is, That the Wood they are made of be well-feafon'd Stuff.

That the Partitions be ftrong, and true let into one another, and that the ends fill up and ftand firm in the Grooves of the *Frame* and middle *Rail* of the *Cafe*.

There is an inconvenience that often happens, thefe thin Partitions, efpecially if they be made of unfeafon'd Stuff, *viz.* as the Stuff dries it fhrinks in the
Grooves

2*

Grooves of the *Frame*, and so not only grows loose, but sometimes starts out above the top of the *Frame*. To prevent this inconvenience, I have of late caused the ends of these thin partitions to be made Male-Duf-tails, broadeft on the under side, and have them fitted into Female-Duf-tails in the Frame of the *Case*, and middle Rail before the bottom Boards are nailed on.

That the Partitions be full an *English* Body thick.

That the Partitions lye close to the bottom of the *Case*, that so the *Letters* slide not through an upper into an under *Box*, when the Papers of the *Boxes* may be worn.

§. 4. *Of* Frames *to set the* Cafes *on.*

Frames are in most *Printing-Houfes* made of thick *Deal-board Battens,* having their several *Rails Tennant-ed* into the *Stiles*: but these sorts of *Frames* are, in respect of their matter (*viz. Fir*) so weak, and in respect of their substance (*viz.* little above an Inch thick) so slight, that experience teaches us, when they are even new made, they tremble and totter, and having lasted a little while, the thinness of their *Tennants* being a little above a quarter of an Inch thick, according to the Rules of *Joynery*, as I have shewn in *Numb.* 5. §. 17. They Craze, their *Tennants* break, or *Morteffes* split, and put the *Mafter-Printer* to a fresh Charge.

It is rationally to be imagined that the *Frames* should be designed to last as long as the *Printing houfe*; and therefore our *Mafter-Printer* ought to take care that they be made of matter strong enough, and of sub-stance big enough to do the Service they are intended for;

for; that they ſtand ſubſtantial and firm in their place, ſo as a ſmall Joſtle againſt them ſhake them not, which often reiterated weakens the *Frame-work*, and at that preſent is ſubject to ſhake the *Letter* in the *Galley* down.

I ſhall not offer to impoſe Rules upon any here, eſpecially ſince I have no Authority from Preſcript or Cuſtom; yet I ſhall ſet down the Scantlings that I my ſelf thought fit to uſe on this occaſion. A De-lineation of the *Frames* are in Plate 1. at C.

a a a a The *Fore-Rails.*
b b b b The *Hind-Rails.*
c The *Top Fore-Rail.*
d The *Bottom-Fore-Rail.*
e The *Top Hind-Rail.*
f The *Bottom Hind-Rail.*
g g g g The *End-Rail.*
h h h h *Croſs-Bearers.*

I made the *Rails* and *Stiles* of well-ſeaſoned fine *Oak*, clean, (that is free from Knots and Shakes) the *Stiles* and *Rails* two Inches and an half ſquare, the Top and Bottom *Fore-Rails* and the Bottom *Hind-Rail* four Foot three Inches long, beſides their *Ten-nants*; And the Top *Hind Rail* five Foot three Inches long. The two *Fore-Rails* and Bottom *Hind-Rail* had Iron Female Screws let into them, which, through an hole made in the *Stiles*, received a Male-Screw with a long ſhank, and a Sholder at the end of it to ſcrew them tight and firm together, even as the Rails of a *Bedſted* are ſcrewed into the *Morteſſes* of a *Bed-Poſt.*   Each

Each *Back Stile* was four Foot one Inch and an half
high befides their *Tennants,* and each *Fore-Stile* three
Foot three Inches high, each *Fore* and *Back-Stile* had
two *Rails* one Foot feven Inches long, befides their
Tennants Tennanted and Pin'd into them, becaufe not
intended to be taken affunder.

It muft be confidered, that the *Fore ftiles* be of a
convenient height for the pitch of an ordinary Man
to ftand and work at, which the heighth aforefaid is;
And that the *Hind ftiles* be fo much higher than the
Fore-ftiles, that when the *Crofs-Bearers* are laid upon
the upper *Fore* and *Hind-Rail,* and the *Cafes* laid on
them, the *Cafes* may have a convenient declivity from
the upper fide the *Upper-Cafe,* to the lower fide the
Lower-Cafe.

The Reafon of this declivity is, becaufe the *Cafes*
ftanding thus before the Workman, the farther *Boxes*
of the *Upper-Cafe* are more ready and eafie to come
at, than if they lay flat; they being in this pofition
fomewhat nearer the hand, and the *Letters* in thofe
Boxes fomewhat eafier feen.

If the Workman prove taller than Ordinary, he
lays another or two pair of *Cafes* under the *Cafes* he
ufes, to mount them: If the Workman be fhort, as
Lads, *&c.* He lays a *Paper-board* (or fometimes two)
on the floor by the Fore-fide of the *Frame,* and ftanding to work on it, mounts himfelf.

The *Bearers* are made of *Slit-Deal,* about two Inches
broad, and fo long as to reach from the *Fore-Rail*
through the *Upper-Rail,* and are let in, fo as to lye
even with the fuperficies of the *Fore* and *Hind-Rail,*
and at fuch a diftance on both the *Rails,* as you may
fee in the Figure.   On

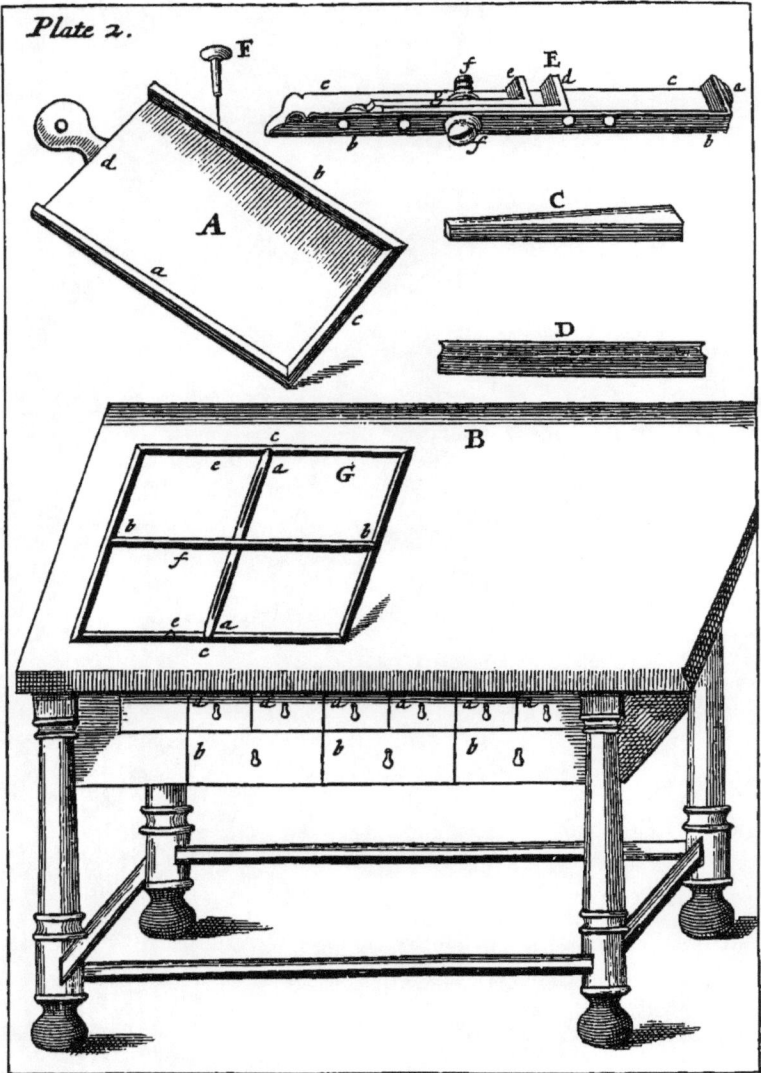

Plate 2.

On the Superficies of the *Fore-Rail,* even with its
Fore-Edge is nailed a fmall *Riglet* about half an Inch
high, and a quarter and half quarter of an Inch
thick, that the *Cafes* fet on the *Frame* having the
aforefaid declivity, may by it be ftop't from fliding
off.

§. 5. *Of the* Galley.

Our *Mafter-Printer* is alfo to provide *Galleys* of different fizes, That the *Compofiter* may be fuited with
fmall ones when he *Compofes* fmall *Pages,* and with
great ones for great *Pages.*
The *Galley* is marked A in Plate 2.

a b c The Sides or Frame of the *Galley.*
d The *Slice.*

Thefe *Galleys* are commonly made of two flat
Wainfcot Boards, each about a quarter and half quarter of an Inch thick, the uppermoft to flide in
Grooves of the Frame, clofe down to the undermoft,
though for fmall *Pages* a fingle Board with two fides
for the Frame may ferve well enough: Thofe *Wainfcot Boards* are an Oblong Square, having its length
longer than its breadth, even as the form of a *Page*
hath.   The three Sides of the Frame are fixed faft
and fquare down on the upper Plain of the undermoft Board, to ftand about three fifth parts of the
height of the *Letter* above the fuperficies of the *Slice.*
The Sides of the Frame muft be broad enough to admit of a pretty many good ftrong *Oaken Pins* along
                                                    the

the Sides, to be drove hard into the Bottom Board, and almoſt quite through the Sides of the Frame, that the Frame may be firmly fixed to it: But by no means muſt they be Glewed on to the Bottom Board, becauſe the *Compoſiter* may ſometimes have occaſion to wet the *Page* in the *Galley*, and then (the *Galley* ſtanding aſlope upon the *Caſe*) the Water will ſoak between the ſides of the Frame, and under Board, and quickly looſen it.

§. 6. *Of the* Correcting-ſtone.

The *Correcting-Stone* marked B in Plate 2. is made of *Marble, Purbeck,* or any other Stone that may be made flat and ſmooth: But yet the harder the Stone is the better; wherefore *Marble* is more preferable than *Purbeck.* Firſt, Becauſe it is a more compact Stone, having fewer and ſmaller Pores in it than *Purbeck.* And Secondly, becauſe it is harder, and therefore leſs ſubject to be prick'd with the corners of a *Chaſe,* if through careleſneſs (as it ſometimes happens) it be pitch'd on the Face of the Stone.

It is neceſſary to have it capacious, *viz.* large enough to hold two *Chaſes* and more, that the *Compoſiter* may ſometimes for his convenience, ſet ſome *Pages* by on it ready to *Impoſe,* though two *Chaſes* lye on the *Stone*: Therefore a *Stone* of about Four Foot and an half long, and Two Foot broad is a convenient ſize for the generality of Work.

This *Stone* is to be laid upon a ſtrong *Oaken*-wood Frame, made like the Frame of a common Table, ſo high, that the Face of the *Stone* may lye about three
Foot

Foot and an Inch above the Floor: And under the
upper Rail of the Frame may be fitted a Row or
two of Draw-Boxes, as at *a a a a a a* and *b b b* on
each of·its longeſt Sides to hold *Flowers, Braſs-Rules,
Braces, Quotations,* ſmall *Scabbords,* &c.

§. 7. *Of* Letter-Boards, *and* Paper-Boards.

Letter-Boards are Oblong Squares, about two
Foot long, eighteen Inches broad, and an Inch and a
quarter thick.   They ought to be made of clean and
well-ſeaſon'd Stuff, and all of one piece: Their upper-
ſide is to be Plained very flat and ſmooth, and their
under-ſide is Clamped with pieces about two Inches
ſquare, and within about four Inches of either end,
as well to keep them from Warping, as to bear them
off the Ground or any other Flat they ſtand on, that
the Fingers of the *Compoſiter* may come at the bottom of the Board to remove it whither he will:
They are commonly made of *Fir*, though not ſo
thick as I have mentioned, or all of one Piece: *Deal-
Boards* of this breadth may ſerve to make them of;
but *Joyners* commonly put *Maſter-Printers* off with
ordinary *Deal-Boards*, which not being broad enough,
they joyn two together; for which cauſe they frequently ſhrink, ſo as the joynt comes aſſunder, and
the *Board* becomes uſeleſs, unleſs it be to ſerve for a
Paper-Board afterwards: For ſmall and thin *Letters*
will, when the Form is open, drop through, ſo as
the *Compoſiter* cannot uſe the Board.

I us'd to make them of *Sugar-Cheſt*; That Stuff being commonly well-ſeaſon'd, by the long lying of the
Sugar

Sugar in it, and is befides a fine hard Wood, and therefore lefs fubject to be injured by the end of the *Shooting-Stick* when a *Form* is *Unlocking.*

Paper-Boards are made juft like the *Letter-Boards,* though feldom fo large, unlefs for great Work: Nor need fuch ftrict care be taken in making them fo exactly fmooth: their Office being only to fet *Heaps of Paper* on, and to *Prefs* the *Paper* with.

§. 8. *Of* Furniture, Quoyns, Scabbord, &c.

By *Furniture* is meant the *Head-fticks, Foot-fticks, Side-fticks, Gutter-fticks, Riglets, Scabbords* and *Quoyns.*

Head fticks and all other *Furniture,* except *Scabbord,* are made of dry *Wainfcot,* that they may not fhrink when the *Form* ftands by; They are *Quadrat* high, ftraight, and of an equal thicknefs all the length: They are made of feveral thickneffes for feveral Works, *viz.* from a *Brevier* which ferves for fome *Quarto's* to fix or eight *Pica* thick, which is many times us'd to *Folio's*: And many of the *Head-fticks* may alfo ferve to make Inner *Side-fticks* of; for the *MafterPrinter* provides them of lengths long enough for the *Compofiter* to cut to convenient Scantlins or Lengths, they being commonly about a Yard long when they come from the *Joyners.* And *Note,* that the *Head* and *Side-fticks* are called *Riglets,* if they exceed not an *Englifh* thick.

Outer *Side-fticks* and *Foot-fticks* marked C in Plate 2. are of the fame heighth of the *Head-fticks, viz. Quadrat* high, and are by the *Joyner* cut to the given length, and to the breadth of the particular *Pages*
that

that are to be *Impofed*: The *Side-fticks* are placed a-
gainft the outer fide of the *Page*, and the *Foot-fticks*
againft the foot or bottom of the *Page*: The outer
fides of thefe *Side* and *Foot-fticks* are bevil'd or floped
from the further to the hither end.

Gutter-fticks marked D in Plate 2. are as the former, *Quadrat* high, and are ufed to fet between *Pages* on either fide the *Croffes*, as in *Octavo's, Twelves,
Sixteens,* and *Forms* upwards; They are made of
an equal thicknefs their whole length, like *Head-
fticks*; but they have a Groove, or Gutter laid on the
upper fide of them, as well that the Water may
drain away when the *Form* is Wafhed or Rinced, as
that they fhould not *Print*, when through the ten-
dernefs of the *Tinpan*, the *Plattin* preffes it and the
Paper lower than ordinary.

Scabbord is that fort of *Scale* commonly fold by
fome *Iron-mongers* in Bundles; And of which, the
Scabbords for *Swords* are made: The *Compofiter* cuts
it *Quadrat* high, and to his Length.

The *Mafter-Printer* is to provide both *Thick* and
Thin Scabbord, that the *Compofiter* may ufe either when
different Bodied *Letter* happens in a *Page*, to juftifie
the *Page* to a true length; And alfo that the *Prefs-
man* may chufe *Thick* or *Thin* to make truer *Regifter*,
as fhall be fhewed in proper place.

Quoyns are alfo *Quadrat* high, and have one of
their fides Bevil'd away to comply with the Bevil
of the *Side* and *Foot-fticks*; they are of different
Lengths, and different Breadths: The great *Quoyns*
about three Inches fquare, except the Bevil on one
fide as aforefaid; and thefe fizes deminifh downwards

to

to an Inch and an half in length, and half an Inch in breadth.

Of these *Quoyns* our *Master-Printer* provides several hundreds, and should provide them of at the least ten different Breadths between the aforesaid sizes, that the *Compositer* may chuse such as will best fit the *Chase* and *Furniture*.

The Office of these *Quoyns* are to *Lock* up the *Form*, *viz.* to wedge it up (by force of a *Mallet* and *Shooting-stick*) so close together, both on the sides and between Head and Foot of the *Page*, that every *Letter* bearing hard against every next *Letter*, the whole *Form* may *Rise*; as shall be shewed hereafter.

Their farther Office is to make *Register* at the *Press*.

§. 9. ¶. 1. *Of the* Mallet, Shooting-stick *and* Dressing-Block, Composing-sticks, Bodkin, *and* Chase. *&c.*

Printers Mallets have a *Cilindrick* Head, and a round Handle; The Head somewhat bigger, and the Handle somewhat longer than those *Joyners* commonly use; Yet neither shape or size different for any reason to be given: But only a Custom always used to have them so. The Head is commonly made of *Beech*.

¶. 2. *Of*

¶. 2. *Of the* Shooting-ftick.

The *Shooting-ftick* muft be made of *Box*, which Wood being very hard, and withal tough, will beft and longeft endure the knocking againft the *Quoyns*. Its fhape is a perfect Wedge about fix Inches long, and its thicker end two Inches broad, and an Inch and an half thick; and its thin end about an Inch and an half broad, and half an Inch thick.

¶. 3. *Of the* Dreffing-Block.

The *Dreffing-Block* fhould be made of *Pear-tree*, Becaufe it is a foft wood, and therefore lefs fubject to injure the Face of the *Letter*; it is commonly a-bout three Inches fquare, and an Inch high. Its Of-fice is to run over the Face of the *Form*, and whilft it is thus running over, to be gently knock't upon with the Head of the *Shooting ftick*, that fuch *Letters* as may chance to ftand up higher than the reft may be preffed down.

Our *Mafter-Printer* muft alfo provide a pair of *Sheers*, fuch as *Taylors* ufe, for the cutting of *Brafs-Rules*, *Scabbords*, &c.

A large *Spunge* or two, or more, he muft alfo provide, one for the *Compofiters* ufe, and for every *Prefs* one.

Pretty fine *Packthread* to tye up *Pages* with; But this is often chofen (or at leaft directed) by the *Com-pofiter*, either finer or courfer, according to the great or fmall *Letter* he works upon.

¶. 4. *Of*

¶. 4. *Of the* Compoſing-ſtick.

Though every *Compoſiter* by Cuſtom is to provide
himſelf a *Compoſing-ſtick*, yet our *Maſter-Printer* ought
to furniſh his Houſe with theſe Tools alſo, and ſuch a
number of them as is ſuitable to the ſize of his Houſe;
Becauſe we will ſuppoſe our *Maſter-Printer* intends to
keep ſome Apprentices, and they, unleſs by contract
or courteſie, are not uſed to provide themſelves *Compoſing-ſticks*: And beſides, when ſeveral *Compoſiters* work
upon the ſame Book, their Meaſures are all ſet alike,
and their *Titles* by reaſon of *Notes* or *Quotations*
broader than their common Meaſure, So that a *Compoſing-ſtick* is kept on purpoſe for the *Titles*, which
muſt therefore be common to all the *Compoſiters* that
work upon that Work; And no one of them is obliged to provide a *Compoſing-ſtick* in common for them
all: Therefore it becomes our *Maſter-Printers* task
to provide them.

It is delineated in Plate 2. at *E*.

a The *Head*.
b b The *Bottom*.
c c The *Back*.
d The lower *Sliding-Meaſure*, or *Cheek*.
e The upper *Sliding-Meaſure*, or *Cheek*.
f f The *Male-Screw*.
g The *Female-Screw*.

Theſe *Compoſing-ſticks* are made of Iron Plate a-
bout the thickneſs of a thin *Scabbord*, and about ten
Inches

Inches long doubled up fquare; fo as the Bottom
may be half an Inch and half a quarter broad,
and the Back about an whole Inch broad.   On the
further end of this Iron Plate thus doubled up, as
at *a* is Soldered on an Iron Head about a *Long-
Primmer* thick; But hath all its outer edges Bafil'd
and Fil'd away into a Molding: This Iron Head muft
be fo let into the Plate, and Soldered on to it, that
it may ftand truly fquare with the bottom, and alfo
truly fquare with the Back, which may be known
by applying the outer fides of a fquare to the Back
and Bottom; as I fhewed, *Numb.* 3. *Fol.* 38, 39.
About two Inches from the Head, in the Bottom, is
begun a row of round holes about an Inch affunder,
to receive the fhank of the *Male-Screw* that fcrews
the *Sliding Meafures* faft down to the Bottom; fo that
the *Sliding-Meafures* may be fet nearer or further
from the Head, as the Meafure of a *Page* may require.

The lower *Sliding-Meafure* marked *d* is an Iron
Plate a *thick Scabbord* thick, and of the Breadth of
the infide of the Bottom; It is about four Inches
long, and in its middle hath a Groove through it
within half an Inch of the Fore-end, and three quarters of an Inch of the hinder end.   This Groove is fo
wide all the way, that it may receive the Shank of the
Screw. On the Fore-end of this Plate ftands fquare upright another Iron Head about a *Brevier* thick, and
reaches fo high as the top of the Back.

The upper *Sliding-Meafure* is made juft like the
lower, only it is about three quarters of an Inch
fhorter.

Between

3

Between thefe two *Sliding-Meafures, Marginal Notes* are *Compofed* to any Width.

Compofiters commonly examine the Truth of their *Stick* by applying the head of the *Sliding-Meafure* to the infide of the Head of the *Stick*; and if they comply, they think they are fquare and true made : But this Rule only holds when the Head it felf is fquare. But if it be not, 'tis eafy to file the *Sliding-Meafures* to comply with them : Therefore, as aforefaid, the fquare is the only way to examine them by.

¶. 5. *Of the* Bodkin.

The *Bodkin* is delineated in Plate 2. at *F* Its *Blade* is made of *Steel,* and well tempered, its fhape is round, and ftands about two Inches without the *Shank* of the *Handle.* The *Handle* is turned of foft wood as *Alder, Maple, &c.* that when *Compofiters* knock the Head of the *Bodkin* upon the Face of a Single *Letter* when it ftands too high, it may not batter the Face.

¶. 6. *Of* Chafes, *marked* G *on the* Correcting-Stone, *Plate* 2.

A *Chafe* is an Iron Frame about two and twenty Inches long, eighteen Inches broad, and half Inch half quarter thick ; and the breadth of Iron on every fide is three quarters of an Inch : But an whole Inch is much better, becaufe ftronger. All its fides muft ftand exactly fquare to each other ; And when it is laid on the *Correcting-Stone* it muft lye exactly flat, *viz.*

viz. equally bearing on all its fides and Angles: The outfide and infide muft be Filed ftraight and fmooth. It hath two *Croffes* belonging to it, *viz.* A *Short-Crofs* marked *a a* and a *Long-Crofs* marked *b b*: Thefe two *Croffes* have on each end a Male Duftail Filed Bevil away from the under to the upper fide of the *Crofs*, fo that the under fide of the Duftail is narrower than the upper fide of the Duftail. Thefe Male-Duftails are fitted into Female-Duftails, Filed in the infide of the *Chafe*, which are alfo wider on the upper fide of the *Chafe* than on the under fide; becaufe the upper fide of the *Crofs* fhould not fall through the lower fide. Thefe *Croffes* are called the *Short* and the *Long Crofs*.

The *Short-Crofs* is Duftail'd in as aforefaid, juft in the middle of the *Chafe* as at *c c*, and the *Long-Crofs* in the middle of the other fides the *Chafe*, as at *d d*. The *Short-Crofs* is alfo Duftail'd into Female-Duftails, made as aforefaid, about three Inches and an half from the middle, as at *e e*: So that the *Short Crofs* may be put into either of the Female-Duftails as occafion ferves. The middle of thefe two *Croffes* are Filed or notched half way through, one on its upper, the other on its under fide to let into one another, *viz.* the *Short-Crofs* is Filed from the upper towards the under fide half way, and the *Long-Crofs* is Filed from the lower towards the upper-fide half way: The *Croffes* are alfo thus let into each other, where they meet at *f*, when the *Short-Crofs* is laid into the other Female-Duftails fitted to it at *e e*.

In the middle, between the two edges of the upper fide of the *Short-Crofs*, is made two Grooves parallel

rallel to the fides of the *Crofs*, beginning at about two Inches from each end, and ending at about feven Inches from each end: It is made about half an Inch deep all the way, and about a quarter of an Inch broad, that the *Points* may fall into them. The *Short-Crofs* is about three quarters of an Inch thick, and the *Long Crofs* about half that thicknefs. All their fides muft be Fil'd ftraight and fmooth, and they muft be all the way of an equal thicknefs.

Hitherto our *Mafter-Printer* hath provided Materials and Implements only for the *Compofiters* ufe; But he muft provide Machines and Tools for the *Prefsmans* to ufe too: which (becaufe I am loath to difcourage my Cuftomers with a fwelling price at the firft reviving of thefe Papers) I fhall (though againft my intereft) leave for the fubject of the next fucceeding *Exercifes*.

ADVERTISEMENTS.

T H E *firft Volumne of* Mechanick Exercifes, *Treating of the* Smiths, *the* Joyners, *the* Carpenters, *and the* Turners *Trades, containing* 37½ *fheets, and* 18 *Copper Cuts, are to be had by the Author.* Jofeph Moxon. *Price* 9 s. 3 d. *in Quires.*

T He *firft Volumne of the Monthly Collection of Letters for Improvement of Husbandry and Trade, containing Twenty four Sheets with an Index, is now finifhed, and the fecond is carrying on:*
By John Haughton, *Fellow of the* Royal Society.

3*

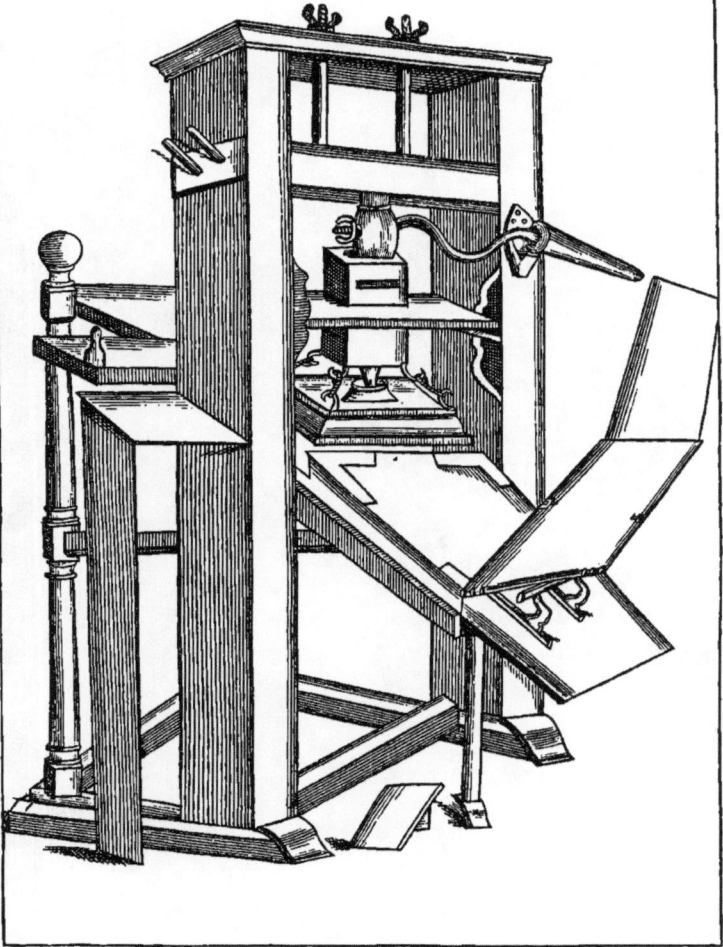

Plate 3.

MECHANICK EXERCISES:

Or, the Doctrine of

Handy-works.

Applied to the Art of

Printing.

§. 10. *Of the* Press.

There are two forts of *Presses* in use, *viz.* the old fashion and the new fashion; The old fashion is generally used here in *England*; but I think for no other reason, than because many *Press-men* have scarce Reason enough to distinguish between an excellently improved Invention, and a make-shift slovenly contrivance, practiced in the minority of this Art.

The New-fashion'd · *Presses* are used generally throughout all the *Low-Countries*; yet because the
Old-

Old-fashion'd *Presses* are used here in *England* (and for no other Reason) I have in Plate 3. given you a delineation of them; But though I give you a draft of them; yet the demensions of every particular Member I shall omit, referring those that think it worth their while, to the *Joyners* and *Smiths* that work to *Printers*: But I shall give a full description of the New-fashion'd *Press*, because it is not well known here in *England*; and if possible, I would for Publick benefit introduce it.

But before I proceed, I think it not amiss to let you know who was the Inventer of this New-fashion'd *Press*, accounting my self so much oblig'd to his Ingeniety for the curiosity of this contrivance, that should I pass by this oppertunity without nameing him, I should be injurious to his Memory.

It was *Willem Jansen Blaew* of *Amsterdam*: a Man as well famous for good and great *Printing,* as for his many *Astronomical* and *Geographical* exhibitions to the World.   In his Youth he was bred up to *Joynery,* and having learn'd his Trade, betook himself (according to the mode of *Holland*) to Travel, and his fortune leading him to *Denmark,* when the noble *Tycho Brahe* was about setting up his *Astronomical Observatory,* was entertain'd into his service for the making his Mathematical-Instruments to Observe withal; in which Instrument-making he shew'd himself so intelligent and curious, that according to the general report of many of his personal acquaintance, all or most of the *Syderal Observations* set forth in *Tycho's* name, he was intrusted to make, as well as the Instruments.

And

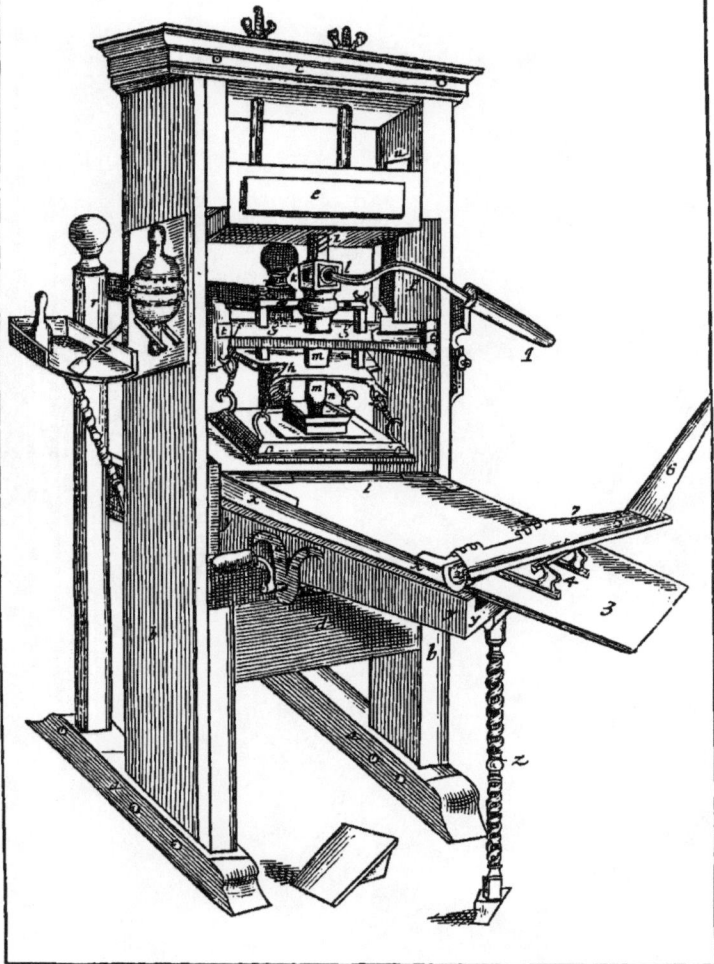

Plate 4.

And before thefe Obfervations were publifh'd to
the World, *Tycho,* to gratify *Blaew,* gave him the
Copies of them, with which he came away to *Am-
fterdam,* and betook himfelf to the making of *Globes,* according to thofe Obfervations.   But as his Trade in-
creafed, he found it neceffary to deal in *Geographical
Maps* and *Books* alfo, and grew fo curious in *Engra-
ving,* that many of his beft *Globes* and *Maps* were *En-
graved* by his own Hands; and by his converfation in
Printing of Books at other *Printing-houfes,* got fuch
in-fight in this Art, that he fet up a *Printing-houfe* of
his own.   And now finding inconveniencies in the
obfolete Invention of the *Prefs,* He contrived a remedy to every inconvenience, and fabricated nine
of thefe New-fafhioned *Preffes,* fet them all on a
row in his *Printing-houfe,* and call'd each *Prefs* by the
name of one of the *Mufes.*

This fhort Hiftory of this excellent Man is, I
confefs forraign to my Title; But I hope my Reader
will excufe the digreffion, confidering it tends only
to the commemoration of a Perfon that hath defer-
ved well of Pofterity, and whofe worth without
this fmall Monument, might elfe perhaps have flid
into Oblivion.

The *Prefs* is a Machine confifting of many Members; it is delineated in Plate 4.

a a The *Feet.*
b b The *Cheeks.*
c The *Cap.*
d The *Winter.*
e The *Head.*
f The *Till.*

                              g g The

g g The *Hofe.* In the Crofs-Iron of which, en-
compaffing the *Spindle,* is the *Garter.*

h h h h The *Hooks* on the *Hofe* the *Plattin* hangs on.

i k l m n The *Spindle.*

i Part of the *Worm* below the *Head,* whofe upper part lies in the *Nut* in the *Head.*

k l The *Eye* of the *Spindle.*

m The *Shank* of the *Spindle.*

n The *Toe* of the *Spindle.*

o o o o The *Plattin* tyed on the *Hooks* of the *Hofe.*

p The *Bar.*

q The *Handle* of the *Bar.*

r r The *Hind-Pofts.*

s s The *Hind-Rails.*

t t The *Wedges* of the *Till.*

u u The *Morteffes* of the *Cheeks,* in which the *Tennants* of the *Head* plays.

x x x x y y The *Carriage.*

x x x x The outer *Frame* of the *Carriage.*

y y The *Wooden-Ribs* on which the *Iron-Ribs* are
faftned.

z The *Stay* of the *Carriage,* or the *Stay.*

1. The *Coffin.*
2. The *Gutter.*
3. The *Planck.*
4. The *Gallows.*
5. The *Tinpans.*
6. The *Frisket.*
7. The *Points.*
8. The *Point-Screws.*

All thefe feveral Members, by their Matter, Form
and Pofition, do particularly contribute fuch an af-
fiftance

fiftance to the whole Machine, that it becomes an Engine managable and proper for its intended purpofe.

But becaufe the fmallnefs of this altogether-Draft may obfcure the plain appearance of many of thefe Parts; Therefore I fhall give you a more particular defcription, and large delineation of every Member in the *Prefs*: And firft of the Wooden work: Where, *Note*, that all the Fram'd Wooden-work of a *Prefs* is made of Good, Fine, Clean, Well-feafon'd *Oak*.

¶. 1. *Of the* Feet.

The *Feet* (marked *a a* in Plate 5.) are two Foot nine Inches and an half long, five Inches deep; and fix Inches broad, and have their out-fides Tryed to a true fquare, as was taught, *Numb.* 5. §. 15. It hath (for ornament fake) its two ends bevil'd away in a Molding, from its upper-fide to its lower, about four Inches within the ends; about four Inches and three quarters within each end of each Foot is made in the middle of the Breadth of the upper-fide of the Foot a Mortefs two Inches wide, to receive the *Tennants* of the lower-end of the *Cheek*, and the *Tennant* of the lower end of the *Hind-Poft*: The Mortefs for the *Cheek* is eight Inches long, *viz.* the Breadth of the *Cheek*: And the Mortefs for the Hind-Poft is four Inches long, *viz.* the fquare of the *Hind-Poft*.

¶. 2. *Of the* Cheeks.

The *Cheeks* (marked *b b* in Plate 5) are five Foot and ten Inches long (befides the *Tennants* of the top

and

and bottom) eight Inches broad, and four Inches and an half thick. All its Sides are tryed fquare to one a-nother. It hath a *Tennant* at either end, its lower *Tennant* marked *a* to enter the Fore-end of the Foot, runs through the middle of the Breadth of the *Cheek,* which therefore is made to fit the Mortefs in the *Foot,* and is about four Inches long, and therefore reaches within an Inch of the bottom of the *Foot;* But the *Tennant* at the upper end of the *Cheek* marked *a,* is cut a-thwart the breadth of the *Cheek,* and therefore can have but four Inches and an half of Breadth, and its thicknefs is two Inches, Its length is four Inches; fo that it reaches into the Mortefs in the *Cap,* within half an Inch of the Top.

In the lower-end-*Tennant* is two holes bored, with-in an Inch and an half of either fide, and within an Inch and an half of the Sholder, with a three quarter Inch *Augure,* to be pin'd into the *Feet* with an Iron Pin.

In the middle of the upper *Tennant,* and within an Inch and an half of the Sholder, is bored another hole, to Pin the *Tennant* into the *Cap,* alfo with an I-ron Pin.

Between *b c* two Foot and half an Inch, and three Foot feven Inches of the Bottom Sholder of the *Tennant, viz.* from the top of the *Winter* to the un-der Sholder the *Till* refts upon, is cut flat away into the thicknefs of the *Cheek,* three Inches in the Infide of the *Cheek;* fo that in that place the *Cheek* remains but an Inch and an half thick : And the *Cheeks* are thus widened in this place, as well becaufe the Duftail *Ten-nants* of the *Winter* may go in between them, as alfo

fo that the *Carriage* and *Coffin* may be made the wider.

Even with the lower Sholder of this flat cutting-in, is made a Duftail Mortefs as at *d*, to reach eight Inches and an half, *viz.* the depth of the *Winter* below the faid Sholder. This Mortefs is three Inches wide on the infide the *Cheek*, and three Inches deep; But towards the infide the *Cheek*, the Mortefs widens in a ftraight line from the faid three Inches to five Inches, and fo becomes a Duftail Mortefs. Into this Duftail Mortefs is fitted a Duftail *Tennant*, made at each end of the *Winter*.

Two Inches above the aforefaid Cutting-in, is another cutting-in of the fame depth, from the Infide the *Cheek* as at *e*. This cutting-in is but one Inch broad at the farther fide the *Cheek*, and an Inch and a quarter on the hither fide the *Cheek*. The under fide of this Cutting-in, is ftraight through the *Cheek*, *viz.* Square to the fides of the *Cheek*: But the upper fide of this Cutting-in, is not fquare through the *Cheeks*, But (as aforefaid) is one quarter of an Inch higher on the fore-fide the *Cheek* than it is on the further fide; So that a Wedge of an Inch at one end, and an Inch and a quarter at the other end may fill this Cutting-in.

At an Inch within either fide the *Cheek*, and an Inch below this Cutting-in, as at *f f*, is made a fmall Mortefs an Inch and an half wide, to which two *Tennants* muft be fitted at the ends of the *Till*, fo that the *Tennants* of the *Till* being flid in through the Cutting-in aforefaid, may fall into thefe Mortefses, and a Wedge being made fit to the Cutting-in, may prefs upon the *Tennants* of the *Till*, and force it down to keep it fteddy in its place.   Here

Here we fee remains a fquare Sholder or fubftance
of Wood between two Cuttings-in; But the under
corner of this fquare Sholder is for Ornament-fake
Bevil'd away and wrought into an *Ogee*.

At two Inches above the laft Cutting-in, is another Cutting-in, but this Cutting-in goes not quite
through the breadth of the *Cheek*, but ftops at an
Inch and an half within the further fide the *Cheek*;
So that above the *Till* and its *Wedge* is another Sholder or fubftance of Wood, whofe upper Corner is alfo
Bevil'd away, and wrought to a Molding as the former.

The laft Cutting-in is marked *g*, and is eight Inches and a quarter above the Sholder of the *Till*, that
it may eafily contain the depth of the *Head*; The
fubftance remaining is marked *h*. This Cutting-in is
made as deep into the thicknefs of the *Cheek* as the
former Cuttings-in are, *viz*. three Inches; and
the reafon the *Cheek* is cut-in here, is, that the *Cheeks*
may be wide enough in this place to receive the *Head*,
and its *Tennants*, without un-doing the *Cap* and *Winter*.

Juft above this Cutting-in is made a fquare Mortefs in the middle of the *Cheek*, as at *i*, it is eight Inches long, and two Inches and an half wide, for the
Tennant of the *Head* to play in.

Upon the fore-fide of the *Cheek* is (for Ornament
fake) laid a Molding through the whole length of
the *Cheek* (a fquare at the Top and Bottom an Inch
deep excepted) it is laid on the outer fide, and therefore can be but an Inch broad; Becaufe the Cuttings-
in on the infide leaves the fubftance of Stuff but an
Inch

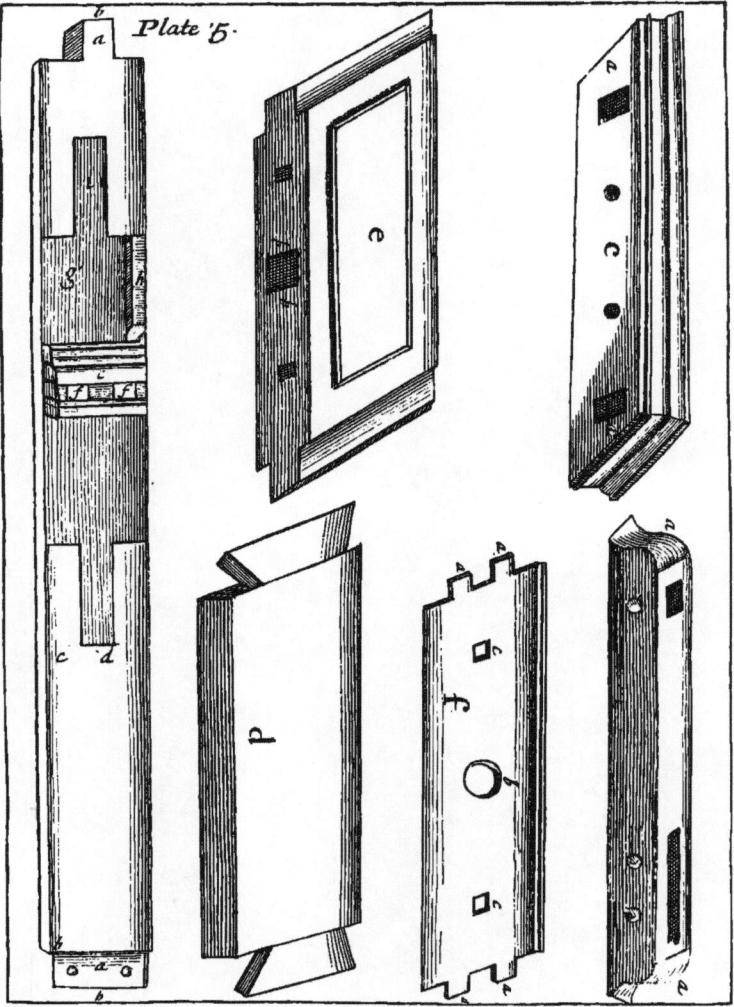

Plate 5.

Inch and an half thick, and fhould the Moldings be
made broader, it would be interrupted in the feveral Cuttings in, or elfe a fquare of a quarter of an
Inch on either fide the Molding could not be allow-
ed, which would be ungraceful.

¶. 3. *Of the* Cap *marked* c *in Plate* 5.

The *Cap* is three Foot and one Inch long, four Inches and an half deep, and nine Inches and an half broad ;
But its fore-fide is cut away underneath to eight Inches, *Viz.* the breadth of the *Cheeks*.   Three quarters
of an Inch above the bottom of the *Cap*, is a fmall
Facia, which ftands even with the thicknefs of the
Cheeks; Half an Inch above that a Bead-Molding,
projecting half an Inch over the *Facia*.   Two Inches above that a broad *Facia*, alfo even with the thicknefs of the *Cheeks*; and an Inch and a quarter above
that is the upper Molding made projecting an Inch
and an half over the two *Facia*'s aforefaid, and the
thicknefs of the *Cheeks*.

Each end of the *Cap* projects three Inches quarter
and half quarter over the *Cheeks*, partly for Ornament, but more efpecially that fubftance may be
left on either end beyond the Morteffes in the *Cap*;
and thefe two ends have the fame Molding laid on
them that the fore-fide of the *Cap* hath.

Within two Inches and half quarter of either end,
on the under-fide the *Cap* is made a fquare Mortefs
two Inches wide, and four Inches and an half long,
viz. the thicknefs of the *Cheek* inwards, as at *a a*, to
receive the Top *Tennants* of the *Cheeks*; which Top

Tennants

Tennants are with an Iron Pin (made tapering of about three quarters of an Inch thick) pin'd into the Mortefs of the *Cap,* to keep the *Cheeks* fteddy in their pofition.

¶. 4. *Of the* Winter *marked* d *in* Plate 5.

The Length of the *Winter* befides the *Tennants,* is one Foot nine Inches and one quarter of an Inch; The Breadth of the *Winter* eight Inches, *viz.* the Breadth of the *Cheek,* and its depth nine Inches; all its fides are tryed fquare; But its two ends hath each a Duftail-*Tennant* made through the whole depth of the *Winter,* to fit and fall into the Duftail Mortefses made in the *Cheeks*: Thefe Duftail-*Tennants* are intended to do the Office of a *Summer,* Becaufe the fpreading of the ends of thefe two *Tennants* into the fpreading of the Mortefses in the *Cheeks,* keeps the two *Cheeks* in a due diftance, and hinders them from flying affunder.

But yet I think it very convenient to have a *Summer* alfo, the more firmly and furer to keep the *Cheeks* together; This *Summer* is only a Rail *Tennanted,* and let into Mortefses made in the infide of the *Cheeks,* and Screwed to them as the Rails defcribed, *Numb.* 15. §. 4. are Screwed into the Stiles of the *Cafe-Frame*; Its depth four Inches and an half, and its breadth eight Inches, *viz,* the breadth of the *Cheeks.*

¶. 5. *Of*

¶. 5. *Of the* Head *marked* e *in* Plate 5.

The length of the *Head* befides the *Tennant* at
either end, is one Foot nine Inches and one quarter
of an Inch; The breadth eight Inches and an half,
and its depth eight Inches. The Top, Bottom and
Hind-fides are tryed Square, but the forefide projects
half an Inch over the Range of the fore-fides of the
Cheeks; in which Projecture is cut a Table with a
hollow Molding about it, two Inches diftant from all
the fides of the fore-fide of the *Head*: Its *Tennants*
are three Inches Broad, and are cut down at either
end, from the top to the bottom of the *Head*,
and made fit to the Morteffes in the *Cheeks*, that they
may flide tight, and yet play in them.
In the under-fide of the *Head* is cut a fquare Hole,
(as at *a*,) about four Inches fquare, and three Inches and
an half deep, into which the *Brafs-Nut* is to be fitted:
And to keep this *Nut* in its place (left the weight of it
fhould make it fall out) is made on either fide the fquare
hole, at about half an Inch diftance from it, (as at *b b*)
a fquare Hole quite through the Top and Bottom of
the *Head* about three quarters of an Inch wide; and
into this fquare Hole is fitted a fquare piece of Iron
to reach quite through the *Head*, having at its under-
end a Hook turned fquare to clafpe upon the under-
fide of the *Nut*; and on its upper-end a Male-Screw
reaching about an Inch above the upper-fide of the
Head, which by the help of a Female-fcrew made
in an Iron *Nut*, with Ears to it to turn it about
draws the *Clafp* at the bottom of the Iron *Shank*
                                                        clofe

4*

clofe againft the *Nut*, and fo keeps it from falling out.

In the middle of the wide fquare Hole that the *Nut* is let into, is bored a round Hole through the top of the *Head*, of about three quarters of an Inch wide, for the *Prefs-man* to pour *Oyl* in at fo oft as the *Nut* and *Spindle* fhall want *Oyling*.

At three Inches from either end of the *Head* (as at *c c*) is bored a Hole quite through the top and bottom of the *Head*, which holes have their under ends fquar'd about two Inches upwards, and thefe fquares are made fo wide as to receive a fquare Bar of Iron three quarters of an Inch fquare; But the other part of thefe Holes remain round: Into thefe Holes two Irons are fitted called the *Screws*.

The Shanks of thefe *Screws* are made fo long as to reach through the *Head* and through the *Cap*: At the upper-end of thefe Shanks is made Male-fcrews, and to thefe Male-fcrews, Iron Female-fcrews are fitted with two Ears to twift them the eafier about.

So much of thefe Iron Shanks as are to lye in the fquare Hole of the *Head* aforefaid, are alfo fquared to fit thofe fquare Holes, that when they are fitted and put into the Holes in the *Head*; they may not twift about.

To the lower-ends of thefe Iron-Shanks are made two Square, Flat Heads, which are let into and buried in the under-fide of the *Head*; And upon the Sholders of thofe two Flat Heads, refts the weight of the *Head* of the *Prefs*; And by the *Screws* at the Upper-end of the Shanks are hung upon the upper-fide of the *Cap*, and Screwed up or let down as occafion requires.

¶. 6. *Of*

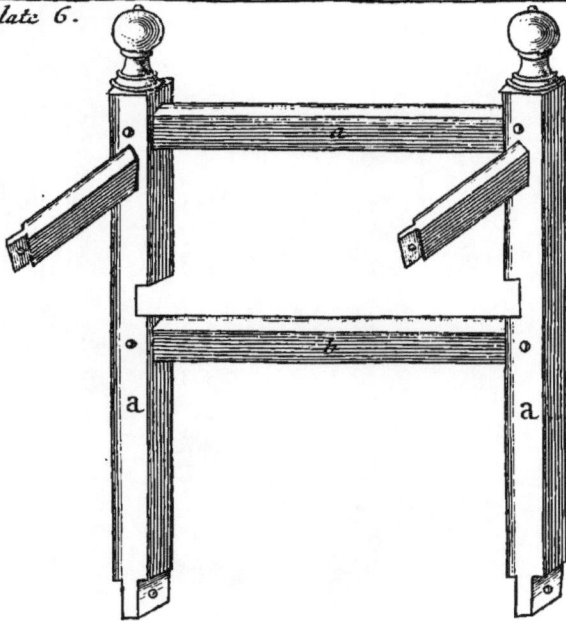

¶. 6. *Of the* Till, *marked* f *in* Plate 5.

The *Till* is a Board about one Inch thick, and is as the *Head* and *Winter*, one Foot nine Inches and a quarter long, befides the *Tennants* at either end; Its Breadth is the Breadth of the *Cheeks, viz.* eight Inches; It hath two *Tennants* at either end as at *a a a a*, each of them about an Inch and an half long, and an Inch and an half broad, and are made at an Inch diftance from the fore and Back-fide, fo that a fpace of two Inches is contained in the middle of the ends between the two *Tennants*; thefe *Tennants* are to be laid in the Morteffes in the *Cheeks* delineated at *f f* in Plate 5. and defcribed in this §. 10. ¶. 2.

In its middle it hath a round Hole about two Inches and an half wide, as at *b*, for the Shank of the *Spindle* to pafs through.

At feven Inches and a quarter from either end, and in the middle between the Fore and Back-fide, is made two fquare Holes through the *Till*, as at *c c*, for the Iron *Hofe* to pafs through.

¶. 7. *Of the* Hind-Pofts *marked* a a *in* Plate 6.

At one Foot diftance from the Hind-fides of the *Cheeks* are placed upright two *Hind-Pofts*, they are three Foot and four Inches long befides the *Tennants*, which *Tennants* are to be placed in the Morteffes in the hinder ends of the *Feet*; Their thicknefs is four Inches on every fide, and every fide is tryed fquare;
But

But within eight Inches of the top is turned a round Ball with a Button on it, and a Neck under it, and under that Neck a ftraight Plinth or Bafe : This turn'd work on the top is only for Ornament fake.

There are fix *Rails* fitted into thefe *Hind-Pofts*, two behind marked *a b*, one of them ftanding with its upper fide at two Inches below the turned Work, the other having its upper-fide lying level with the upper-fide of the *Winter*.

Thefe two *Rails* are each of them *Tennanted* at either end, and are made fo long, that the out-fides of the *Hind-Pofts* may ftand Range or even with the outer-fides of the *Cheeks*; Thefe *Tennants* at either end are let into Mortefles made in the in-fides of the *Hind-Pofts*, and Pin'd up with half Inch wooden Pins, Glewed in, as was fhewn Vol. 1. *Numb. 5. §. 17.* Becaufe the two *Hind-Pofts* need not be feparated for any alteration of the *Prefs*.

The two *Side-Rails* on either fide the *Prefs* are *Tennanted* at each end, and let into Mortefles made in the *Cheeks* and *Hind-Pofts*, fo as they may ftand Range with the outer-fides of the *Cheeks* and *Hind-Pofts*; But the *Tennants* that enter the Mortefles in the *Cheeks* are not pin'd in with Wooden Pins, and Glewed, becaufe they may be taken affunder if need be; But are Pin'd in with Iron Pins, made a little tapering towards the entring end, fo as they may be driven back when occafion ferves to alter the *Prefs*: And the *Tennants* that enter the Mortefles in the *Hind-Pofts* are faftned in by a Female-fcrew, let in near the end of the *Rail*, which receives a Male-fcrew thruft through the *Hind-Pofts*, even as I fhew'd in

§. 4.

§. 4. the *Fore* and *Back-Rails* of the *Cafe-Frames* was.

¶. 8. *Of the* Ribs *marked* b *in* Plate 6.

The *Ribs* lye within a Frame of four Foot five Inches long, one Foot eleven Inches broad; its two *End-Rails* one Inch and an half thick, its *Side-Rails* two Inches and an half thick; and the breadth of the *Side* and *End-Rails* two Inches and an half. But the *Side-Rails* are cut away in the in-fide an Inch and an half towards the outer fides of the *Rails,* and an Inch deep towards the Bottom fides of the *Rails,* fo that a fquare *Cheek* on either *Side-Rail* remains. This cutting down of the *Outer-Rails* of the *Frame* is made, becaufe the Planck of the *Carriage* being but one Foot eight Inches and an half broad, may eafily flide, and yet be gaged between thefe *Cheeks* of the *Rail,* that the *Cramp-Irons* Nailed under the *Carriage Planck* joggle not on either fide off the *Ribs,* as fhall more fully be fhewn in the next §.

Between the two *Side-Rails* are framed into the two *End-Rails* the two *Wooden-Ribs* two Inches and an half broad, and an Inch and an half thick; they are placed each at an equal diftance from each *Side-Rail,* and alfo at the fame diftance between themfelves. Upon thefe two *Ribs* are faft Nailed down the *Iron-Ribs,* of which more shall be faid when I come to fpeak of the Iron-work.

¶. 9. *Of*

¶. 9. *Of the* Carriage, Coffin *and its* Planck,
marked a *in* Plate 7.

The *Planck* of the *Carriage* is an Elm-Planck an
Inch and an half thick, four Foot long, and one Foot
eight Inches and three quarters broad; upon this
Planck at its fore-end is firmly Nailed down a fquare
Frame two Foot four Inches long, one Foot ten
Inches broad, and the thicknefs of its Sides two Inch-
es and an half fquare; This Frame is called the *Coffin*,
and in it the *Stone* is *Bedded*.

Upon each of the four Corners of this *Coffin* is let
in and faftned down a fquare Iron Plate as at *a a a a*,
with Return Sides about fix Inches long each fide,
half a quarter of an Inch thick, and two Inches and
a quarter broad; upon the upper outer-fides of each
of thefe Plates is faftned down to them with two or
three Rivets through each fide, another ftrong Iron
half an Inch deep, and whofe outer Angles only are
fquare, but the Inner Angles are obtufe, as being
floped away from the Inner-Angle towards the far-
ther-end of each inner-fide, fo as the *Quoins* may do
the Office of a Wedge between each inner-fide and
the *Chafe*.

The Plates of thefe Corners (as I faid) are let in
on the outer-Angles of the upper-fide of the Frame
of the *Coffin*, fo as the upper-fides of the Plates lye
even with it, and are Nailed down, or indeed rather Rivetted down through the bottom and top-
fides of the Frame of the *Coffin*, becaufe then the upper-fides of the Holes in the Iron Plates being fquare
Bored

Plate 7.

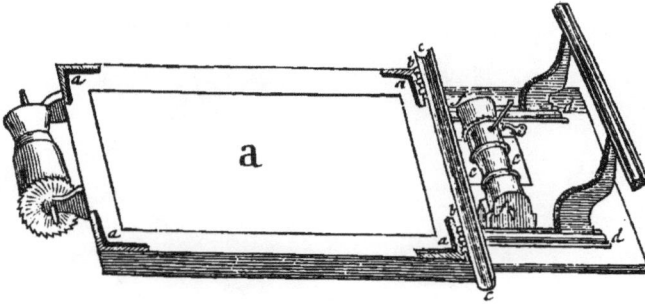

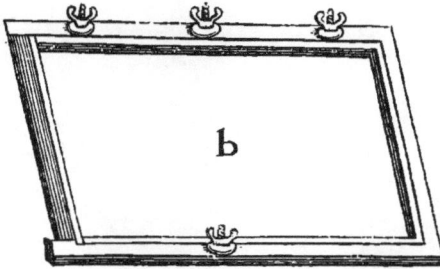

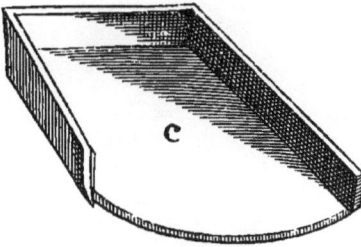

Bored (that is, made wider on the upper fide of the Plate, as I fhall fhew when I come to the making of *Mathematical Inftruments*) the ends of the Shanks of the Iron Pins may be fo battered into the Square-boring, that the whole Superficies of the Plate when thus Rivetted fhall be fmooth, which elfe with the exturberancies of Nail-heads would hinder the free fliding of the *Quoins*.

At the hinder end of the Frame of the *Coffin* are faftned either with ftrong Nails, Rivets, or rather Screws, two Iron *Half-Joynts*, as at *b b*, which having an Iron Pin of almoft half an Inch over put through them, and two *Match-half-Joynts* faftned on the Frame of the *Tympan*, thefe two *Match-half-Joynts* moving upon the Iron Pin aforefaid, as on an *Axis*, keeps the *Tinpan* fo truly gaged, that it always falls down upon the *Form* in the place, and fo keeps *Regifter* good, as fhall further be fhewed in proper place.

Behind the *Coffin* is Nailed on to its outfide, a Quarter, as at *c c* this Quarter is about three Inches longer than the breadth of the *Coffin*, it hath all its fides two Inches over, and three of them fquare; but its upper fide is hollowed round to a Groove or Gutter an Inch and an half over. This Gutter is fo Nailed on, that its hither end ftanding about an Inch higher than its further end, the Water that defcends from the *Tympan* falling into it is carried away on the farther fide the *Coffin* by the declivity of the farther end of the Gutter, and fo keeps the Planck of the *Carriage* neat and cleanly, and preferves it from rotting.

Parallel

Parallel to the outer fides of the hind part of the
Planck of the *Carriage*, at three Inches diftance from
either fide, is Nailed down on the upper fide of the
Planck two Female-Duftail Grooves, into which is
fitted (fo as they may flide) two Male-Duftails made
on the two Feet of the *Gallows* (as at *d d*) that the
Tinpan refts upon; and by the fliding forward or
backward of thefe Duftail Feet, the heighth of the
Tinpan is raifed or depreffed according to the Reafon
or Fancy of the *Prefs-man*.

At three Inches from the hinder Rail of the *Coffin*,
in the middle, between both fides of the Planck, is
cut an Hole four Inches fquare (as at *e e*) and upon the hither and farther fide of this Hole is faftned down on each fide a *Stud* made of Wood (as at
f f) and in the middle of thefe two *Studs* is made a
round Hole about half an Inch over, to receive the
two round ends of an Iron Pin; which Iron Pin,
though its ends be round, is through the middle of
the Shank fquare, and upon that fquare is fitted a
round *Wooden-Rowler* or *Barrel*, with a Shoulder on
either fide it, to contain fo much of the *Girt* as fhall
be rowled upon it; And to one end of the *Rowler* is
faftned an Iron *Circle* or *Wheel*, having on its edge
Teeth cut to ftop againft a *Clicker*, when the *Rowler*
with an Iron Pin is turned about to ftrain the *Girt*.

¶. 10. *Of the* Tympan *and* Inner-Tympan, *marked* b *in* Plate 7.

The *Tympan* is a fquare Frame, three fides whereof
are Wood, and the fourth Iron. Its width is one
Foot

Foot eight Inches, its length two Foot two Inches; the breadth of the wooden Sides an Inch and an half, and the depth one Inch.

On its ſhort Wooden-ſide, *viz.* its Hind-end, at the two Corners is Rivetted an Iron *Match-Joynt*, to be pinned on to another *Half joynt* faſtned on the *Hind-Rail* of the *Coffin.*

The other end, *viz.* the Fore-end of the *Tympan* is made of Iron, with a ſquare *Socket* at either end for the Wooden ends of the *Tympan* to fit and faſten into. This Iron is ſomewhat thinner and narrower than an ordinary Window-Caſement.

Upon the outer edge of this Iron, about an Inch and an half off the ends of it, is made two Iron *Half-joynts* to contain a Pin of about a quarter of an Inch over, which Pin entring this *half-joynt*, and a *match Half-joynt* made upon the *Frisket*, ſerves for the *Frisket* to move truly upon.

In the middle of each long *Rail* of the *Tympan*, is made through the top and bottom an Hole half an Inch ſquare, for the ſquare Shanks of the *Point-Screws* to fit into.

The like Holes are alſo made in the *Tympan*, at one third part of its length from the Fore-end or *Frisket-joynt*, to place the *Point Screws* in; when a *Twelves*, *Eighteens*, &c. is wrought.

Into the Inner-ſide of this *Tympan* is fitted the *Inner-Tympan*, whoſe three ſides are alſo made of Wood, and its fourth ſide of Iron, as the *Tympan*, but without *joynts*; it is made ſo much ſhorter than the *Outer-Tympan*, that the outer edge of the Iron of the *Inner-Tympan* may lye within the inner edge of the Iron on the *Outer-Tympan*;

5

pan; and it is made fo much narrower than the infide
of the *Tympan,* that a convenient room may be allow-
ed to pafte a *Vellom* between the infide of the *Tympan,*
and the outfide of the *Inner-Tympan.*

About the middle, through the hither-fide of the
Inner-Tympan, is let in and faftned an Iron Pin about
a quarter of an Inch over, and ftands out three quarters of an Inch upon the hither out-fide of the *Inner-
Tympan,* which three quarters of an Inch Pin fits into
a round hole made in the inner-fide of the *Tympan,* to
gage and fit the *Inner-Tympan* right into the *Tympan*;
for then by the help of an Iron turning *Clafp* on the
further fide the *Tympan,* the *Inner-Tympan* is kept firmly
down and in its pofition.

¶. 11. *Of the* Inck-Block, Slice, Brayer, *and* Catch *of the* Bar, *marked* c d e f in Plate 7.

To the *Rail* between the hither *Cheek* and *Hind-
Poft* is faftned the *Inck-Block,* which is a Beechen-board
about thirteen Inches long, nine Inches broad, and
commonly about two Inches thick, and hath the
left hand outer corner of it cut away; it is Railed in
on its farther and hinder-fides, and a little above half
the hither-fide, with Wainfcot-Board about three
quarters of an Inch thick, and two Inches and an
half above the upper-fide the board of the *Inck-Block.*
It is defcribed in Plate 7. at c.

The *Brayer* marked *a* is made of *Beech*: It is turned
round on the fides, and flat on the bottom, its length

is

is about three Inches, and its diameter about two Inches and an half; it hath an Handle to it about four Inches long. Its Office is to rub and mingle the *Inck* on the *Inck-Block* well together.

The *Slice* is a little thin Iron *Shovel* about three or four Inches broad, and five Inches long; it hath an Handle to it of about ſeven Inches long. Near the *Shovel* through the Handle is fitted a ſmall Iron of about two Inches long ſtanding Perpendicular to both the ſides of the Handle, and is about the thickneſs of a ſmall Curtain-Rod. It is deſcribed at e.

The *Catch* of the *Bar* deſcribed at f is a piece of Wood two Inches thick, four Inches broad, and ten Inches long; The top of it is a little Bevil'd or Slop'd off, that the *Bar* may by its *Spring* fly up the Bevil till it ſtick. This Bevil projeꞔts three Inches over its ſtraight Shank, which reaches down to the bottom; in the middle of this Shank, through the fore and back-ſide, is a Morteſs made from within an Inch of the Rounding to an Inch and an half of the bottom; This Morteſs is three quarters of an Inch wide, and hath an Iron Pin with a Shoulder at one end fitted to it, ſo as it may ſlide from one end of the Morteſs to the other. At the other end of the Iron Pin is made a Male-Screw which enters into a Female Iron Screw let into the further *Cheek* of the *Preſs*; ſo that the *Catch* may be Screwed cloſe to the *Cheek*, as ſhall further be ſpoken to hereafter.

¶. 12. *Of*

¶. 12. *Of the Iron-work, and firſt of the*
Spindle *marked* A *in Plate* 8.

From the Top to the *Toe* of the *Spindle, viz.* from
a to *b* is ſixteen Inches and a half, the length of the
Cilinder the *Worms* are cut upon is three Inches and a
quarter, and the diameter of that Cilinder two Inches and a quarter; between the bottom of the *Worms*
and top of the Cube one Inch and an half; the Cube
marked *c c c c* is two Inches and three quarters, the
ſquare *Eye* at *d* in the middle of the Cube is an Inch
and a quarter through all the ſides of the Cube; one
Inch under the Cube at *e* is the *Neck* of the *Spindle,*
whoſe diameter is two Inches, It is one Inch between
the two ſhoulders, *viz.* the upper and under ſhoulders of the *Neck* at *e e,* ſo that the Cilinder of the
Neck is one Inch long; the very bottom of the *Spindle* at *b* is called the *Toe,* it is made of an Hemiſphe-
rical form, and about one Inch in diameter; This *Toe*
ſhould be made of *Steel,* and well Temper'd, that
by long or careleſs uſage, the point of preſſure wear
not towards one ſide of the *Toe,* but may remain in
the Axis of the *Spindle.*

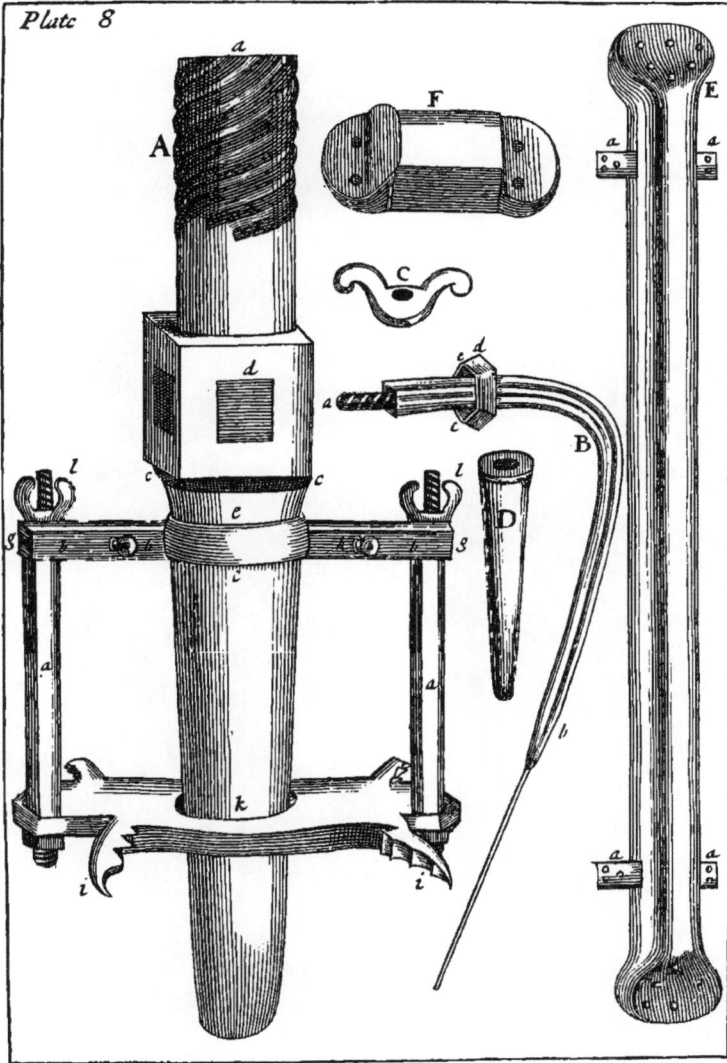

Plate 8

A

a

F

C

d

c   c

e

a

e

l   l

S   b   b   S

a   a

k

D

B

e d

a
c

b

E

a   a

a   a

5*

§. 11. *Of the* Worms *of the* Spindle.

I promifed at the latter end of *Numb.* 2. to give a more copious account than there I did of making *Worms*, when I came to exercife upon *Printing-Prefs Spindles*; and being now arrived to it, I fhall here make good my promife.

¶. 1. The *Worms* for *Printing-Prefs Spindles* muft be projeçted with fuch a declivity, as that they may come down at an affigned progrefs of the *Bar.*

The affigned progrefs may be various, and yet the *Spindle* do its office: For if the *Cheeks* of the *Prefs* ftand wide affunder, the fweep or progrefs of the fame *Bar* will be greater than if they ftand nearer together.

It is confirm'd upon good confideration and Reafon as well as conftant experience, that in a whole Revolution of the *Spindle*, in the *Nut*, the *Toe* does and ought to come down two Inches and an half; but the *Spindle* in work feldom makes above one quarter of a Revolution at one *Pull*, in which fweep it comes down but half an Inch and half a quarter of an Inch; and the reafon to be given for this coming down, is the fqueezing of the feveral parts in the *Prefs*, fubjeçt to fqueeze between the Morteffes of the *Winter* and the Morteffes the *Head* works in; and every Joynt between thefe are fubjeçt to fqueeze by the force of a *Pull.* As firft, The *Winter* may fqueeze down into its Mortefs one third part of the thicknefs of a *Scabbord.* (Allowing a *Scabbord* to be half a *Nomparel* thick.) Secondly, The *Ribs* fqueeze clofer to

the

the *Winter* one *Scabbord.*   Thirdly, The *Iron-Ribs*
to the Wooden *Ribs* one *Scabbord.*   Fourthly, The
Cramp-Irons to the *Planck* of the *Coffin* one *Scabbord.*
Fifthly, The *Planck* it felf half a *Scabbord.*   Sixthly,
The *Stone* to the *Planck* one *Scabbord.*   Seventhly,
The *Form* to the *Stone* half a *Scabbord.*   Eighthly,
The *Juſtifyers* in the Morteſs of the *Head* three *Scab-*
bords.   Ninthly, The *Nut* in the *Head* one *Scabbord.*
Tenthly, The *Paper,* *Tympans* and *Blankets* two
Scabbords.   Eleventhly, Play for the Irons of the *Tym-*
pans four *Scabbords.*   Altogether make fifteen *Scab-*
bords and one third part of a *Scabbord* thick, which
(as aforeſaid) by allowing two *Scabbords* to make a
Nomparel, and as I ſhewed in *Vol.* 2. *Numb.* 2. §. 2.
One hundred and fifty *Nomparels* to make one Foot,
gives twelve and an half *Nomparels* for an Inch, and
conſequently   twenty   five   *Scabbords*   for   an   Inch;
ſo by proportion, fifteen *Scabbords* and one third part
of a *Scabbord,* gives five eighth parts of an Inch,
and a very ſmall matter more, which is juſt ſo much
as the *Toe* of the *Spindle* comes down in a quarter of
a Revolution.

   This is the Reaſon that the coming down of the
Toe ought to be juſt thus much; for ſhould it be leſs,
the natural Spring that all theſe Joynts have, when
they are unſqueez'd, would mount the Irons of the
Tympans ſo high, that it would be troubleſom and tedious for the *Preſs-man* to *Run* them under the *Plat-*
tin, unleſs the *Cheeks* ſtood wider aſſunder, and conſequently every ſweep of the *Bar* in a *Pull* exceed a
quarter of a Revolution, which would be both laborious for the *Preſs-man,* and would hinder his uſual
riddance of Work.   I ſhew'd

I ſhew'd in *Numb.* 2. *fol.* 31, 32, 33, 34, 35. the
manner of making a Screw in general; but aſſigned
it no particular Riſe; which for the aforeſaid reaſon,
theſe *Printing-Preſs Screws* are ſtrictly bound to have:
Therefore its aſſigned Riſe being two Inches and an
half in a Revolution, This meaſure muſt be ſet off
upon the Cilindrick Shank, from the top towards
the Cube of the *Spindle,* on any part of the *Cilinder,*
and there make a ſmall mark with a fine *Prick-Punch,*
and in an exact Perpendicular to this mark make a-
nother ſmall mark on the top of the Cilinder, and
laying a ſtraight *Ruler* on theſe two marks, draw a
ſtraight line through them, and continue that line
almoſt as low as the Cube of the *Spindle.* Then devide that portion of the ſtraight line contained between the two marks into eight equal parts, and ſet
off thoſe equal parts from the two Inch and half
mark upwards, and then downwards in the line ſo
oft as you can: Devide alſo the Circumference of
the Shank of the Cilinder into eight equal parts, and
draw ſtraight lines through each deviſion, parallel to
the firſt upright line; and deſcribe the *Screw* as you
were directed in the afore-quoted place; ſo will you
find that the revolution of every line ſo carried on
about the Shank of the Cilinder, will be juſt two
Inches and an half off the top of the Shank: which
meaſure and manner of working may be continued
downward to within an Inch and an half of the Cube
of the *Spindle.* This is the Rule and Meaſure that
ought to be obſerv'd for ordinary *Preſſes*: But if for
ſome by-reaſons the aforeſaid Meaſure of two Inches
and an half muſt be varied, then the varied Meaſure
muſt

muſt be ſet off from the top of the Cilinder, and
working with that varied Meaſure as hath been directed, the *Toe* of the *Spindle* will come down lower
in a revolution if the varied Meaſure be longer, or
not ſo low if the varied Meaſure be ſhorter.

There is a Notion vulgarly accepted among Work-
men, that the *Spindle* will Riſe more or leſs for the
number of *Worms* winding about the Cilinder; for
they think, or at leaſt by tradition are taught to ſay,
that a *Three-Worm'd Spindle* comes faſter and lower
down than a *four-Worm'd Spindle*: But the opinion
is falſe; for if a *Spindle* were made but with a *Single-
Worm*, and ſhould have this Meaſure, *viz.* Two
Inches and an half ſet off from the top, and a *Worm*
cut to make a Revolution to this Meaſure, it would
come down juſt as faſt, and as low, as if there were
two, three, four, five or ſix *Worms*, &c. cut in the
ſame Meaſure: For indeed, the numbers of *Worms*
are only made to preſerve the *Worms* of the *Spindle*
and *Nut* from wearing each other out the faſter; for
if the whole ſtreſs of a *Pull* ſhould bear againſt the
Sholder of a ſingle *Worm*, it would wear and ſhake
in the *Nut* ſooner by half than if the ſtreſs ſhould
be borne by the Sholders of two *Worms*; and ſo pro-
portionably for three, four, five *Worms*, &c.

But the reaſon why four *Worms* are generally made
upon the *Spindle*, is becauſe the Diameters of the *Spindle* are generally of this propos'd ſize; and therefore a
convenient ſtrength of Mettal may be had on this
ſize for four *Worms*; But ſhould the Diameter of the
Spindle be ſmaller, as they ſometimes are when the
Preſs is deſigned for ſmall Work, only three *Worms*
                                                      will

will be a properer number than four; becaufe when
the Diameter is fmall, the thicknefs of the *Worms*
would alfo prove fmall, and by the ftrefs of a *Pull*
would be more fubject to break or tear the *Worms*
either of the *Spindle* or *Nut*.

And thus I hope I have performed the promife here
I made at the latter end of *Numb*. 2. Whither I refer you for the breadth, and reafon of the breadth of
the *Worm*.

¶. 1 3. *Of the* Bar *marked* B *in* Plate 8.

This *Bar* is Iron, containing in length about two
Foot eight Inches and an half, from *a* to *b*, and its
greateft thicknefs, except the Sholder, an Inch and a
quarter; The end *a* hath a Male-Screw about an Inch
Diameter and an Inch long, to which a *Nut* with a
Female-Screw in it as at C is fitted. The Iron *Nut*
in which this Female-Screw is made, muft be very
ftrong, *viz.* at leaft an Inch thick, and an Inch and
three quarters in Diameter; in two oppofite fides
of it is made two Ears, which muft alfo be very
ftrong, becaufe they muft with heavy blows be knock't
upon to draw the Sholder of the fquare fhank on the
Bar, when the fquare Pin is in the *Eye* of the *Spindle*
clofe and fteddy up to the Cube on the *Spindle*. The
fquare Pin of the *Bar* marked *c* is made to fit juft
into the *Eye*, through the middle of the Cube
of the *Spindle*, on the hither end of this fquare Pin
is made a Sholder or ftop to this fquare Pin, as at *d*.
This Sholder muft be Filed exactly Flat on all its four
in-fides, that they may be drawn clofe and tight up
to

to any flat fide of the Cube on the *Spindle*; It is two Inches fquare, that it may be drawn the firmer, and ftop the fteddyer againft any of the flat fides of the faid Cube, when it is hard drawn by the ftrength of the Female-Screw in the aforefaid *Nut* at C. The thicknefs from *d* to *e* of this Sholder is about three quarters of an Inch, and is Bevil'd off towards the *Handle* of the *Bar* with a fmall Molding.

The fubftance of this *Bar*, as aforefaid, is about an Inch and a quarter; but its Corners are all the way flatted down till within five Inches of the end: And from thefe five Inches to the end, it is taper'd away, that the *Wooden-Handle* may be the ftronger forced and faftned upon it.

About four Inches off the Sholder, the *Bar* is bowed beyond a right Angle, yet not with an Angle, but a Bow, which therefore lies ready to the *Prefs-man*'s Hand, that he may Catch at it to draw the *Wooden-Handle* of the *Bar* within his reach.

This *Wooden-Handle* with long Working grows oft loofe; but then it is with hard blows on the end of it forced on again, which oft fplits the *Wooden-Handle* and loofens the fquare Pin at the other end of the *Bar*, in the *Eye* of the *Spindle*: To remedy which inconvenience, I ufed this Help, *viz.* To weld a piece of a Curtain-Rod as long as the *Wooden-Handle* of the *Bar*, to the end of the Iron *Bar*, and made a Male-fcrew at the other end with a Female-fcrew to fit it; Then I bored an hole quite through the *Wooden-Handle*, and Turn'd the very end of the *Wooden-Handle* with a fmall hollow in it flat at the bottom, and deep enough to bury the Iron-*Nut* on the end of the Cur-tain-

tain-Rod, and when this Curtain-Rod was put through the Hollow in the *Wooden Handle* and Screwed faſt to it at the end, it kept the *Wooden-Handle*, from flying off; Or if it looſened, by twiſting the *Nut* once or twice more about, it was faſtned again.

¶. 14. *Of the* Hoſe, Garter, *and* Hoſe-Hooks.

The *Hoſe* are the upright Irons in Plate 8. at *a a*, They are about three quarters of an Inch ſquare, both their ends have Male-ſcrews on them; The lower end is fitted into a ſquare Hole made at the parting of the *Hoſe-Hooks*, which by a ſquare *Nut* with a Female-ſcrew in it, is Screwed tight up to them; Their upper ends are let into ſquare Holes made at the ends of the *Garter*, and by *Nuts* with Female-Screws in them, and Ears to turn them about as at *l l* are drawn up higher, if the *Plattin-Cords* are too looſe; or elſe let down lower if they are too tight: Theſe upper Screws are called the *Hoſe-Screws*.

The *Garter* (but more properly the *Coller*) marked *b b*, is the round Hoop incompaſſing the flat Groove or Neck in the Shank of the *Spindle* at *e e*; This round Hoop is made of two half round Hoops, having in a Diametrical-line without the Hoop ſquare Irons of the ſame piece proceeding from them, and ſtanding out as far as *g g*, Theſe Irons are ſo let into each other, that they comply and run Range with the ſquare Sholders at both ends, wherein ſquare Holes are made at the ends of the *Hoſe*. They are Screwed together with two ſmall Screws, as at *h h*.

The

The four *Hofe-Hooks* are marked *i i i i,* They proceed from two Branches of an Iron Hoop at *k* en-
compaffing the lower-end of the *Spindle,* on either
Corner of the Branch, and have notches filled in their
outer-fides as in the Figure, which notches are to contain feveral Turns of *Whip-cord* in each notch, which
Whip-cord being alfo faftned to the *Hooks* on the *Plattin,* holds the *Plattin* tight to the *Hooks* of the *Hofe.*

¶. 15. *Of the* Ribs, *and* Cramp-Irons.

The *Ribs* are delineated in Plate 8. at E, they
are made of four-fquare Irons the length of the *Wooden-Ribs* and *End-Rails, viz.* Four Foot five Inches
long, and three quarters of an Inch fquare, only one
end is batter'd to about a quarter of an Inch thick, and
about two Inches and an half broad, in which battering four or five holes are Punch't for the nailing
it down to the *Hind-Rail* of the *Wooden-Ribs.* The
Fore-end is alfo batter'd down as the Hind-end, but
bound downwards to a fquare, that it may be nailed
down on the outer-fide of the *Fore-Rail* of the *Wooden-
Ribs.*

Into the bottom of thefe *Ribs,* within nine Inches
of the middle, on either fide is made two Female-
Duftails about three quarters of an Inch broad, and
half a quarter of an Inch thick, which Female-Duftails have Male-Duftails as at *a a a a* fitted ftiff into
them, about an Inch and three quarters long; and
thefe Male-Duftails have an hole punched at either
end, that when they are fitted into the Female-Duftails in the *Ribs,* they may in thefe Holes be Nailed
down the firmer to the *Wooden-Ribs.*   Thefe

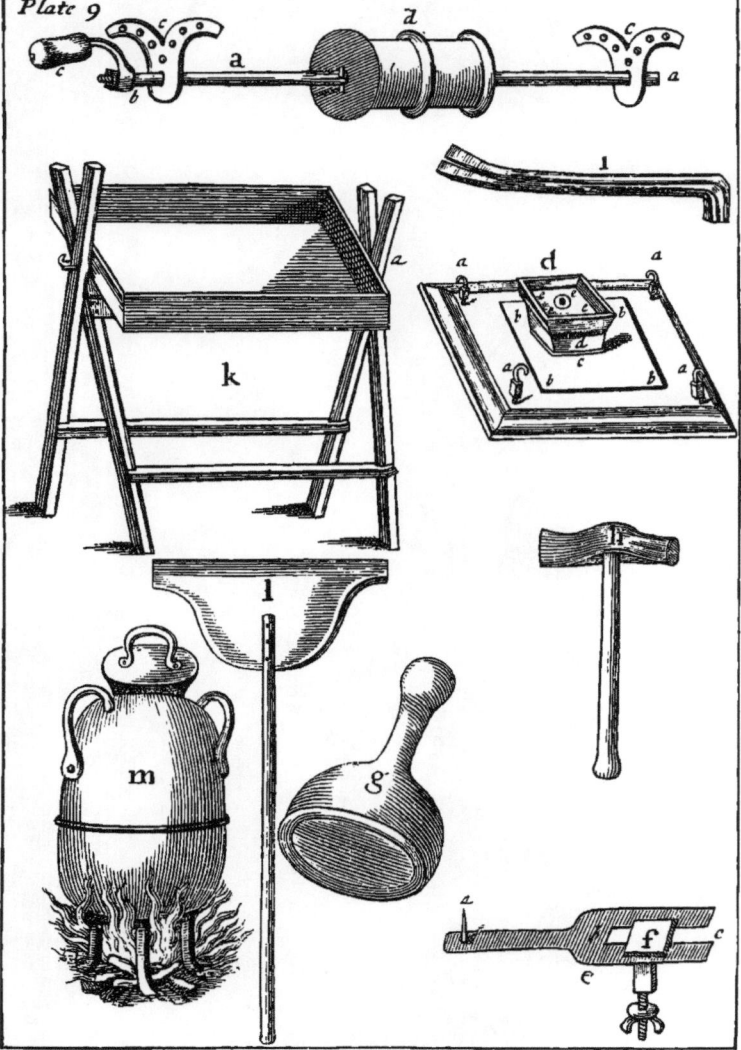

Plate 9

Thefe *Ribs* are to be between the upper and the under fide exactly of an equal thicknefs, and both to lye exactly Horizontal in ftraight lines; For irregularities will·both Mount and Sink the *Cramp-Irons*, and make them *Run* rumbling upon the *Ribs*.

The upper-fides of thefe *Ribs* muft be purely Smooth-fil'd and Pollifh'd, and the edges a little Bevil'd roundifh away, that they may be fomewhat Arching at the top; becaufe then the *Cramp-Irons Run* more eafily and ticklifhly over them.

The *Cramp-Irons* are marked F in Plate 8. They are an Inch and an half long befides the Battering down at both ends as the *Ribs* were; They have three holes Punched in each Battering down, to Nail them to the *Planck* of the *Coffin*; They are about half an Inch deep, and one quarter and an half thick; their upper-fides are fmoothed and rounded away as the *Ribs*.

¶. 16. *Of the* Spindle *for the* Rounce, *defcribed in Plate* 9. *at* a.

The *Axis* or *Spindle* is a ftraight Bar of Iron about three quarters of an Inch fquare, and is about three Inches longer than the whole breadth of the Frame of the *Ribs*, *viz.* two Foot two Inches: The farther end of it is Filed to a round Pin (as at *a*) three quarters of an Inch long, and three quarters of an Inch in Diameter; the hither end is filed away to fuch another round Pin, but is two Inches and a quarter long (as at *b*); at an Inch and a quarter from this end is Filed a Square Pin three quarters of an Inch long, and

with-

within half an Inch of the end is Filed another round
Pin, which hath another Male-Screw on it, to which
is fitted a fquare Iron *Nut* with a Female-Screw in it.

On the Square Pin is fitted a *Winch* fomewhat in
form like a Jack-winch, but much ftronger; the *Eye*
of which is fitted upon the Square aforefaid, and
Screwed up tight with a Female-Screw.   On the
ftraight Shank of this Winch is fitted the *Rounce*,
marked *e*.

The round ends of this *Axis* are hung up in two
Iron-Sockets (as at *c c*) faftned with Nails (but
more properly with Screws) on the outfide the Wood-
en Frame of the *Ribs*.

The *Girt-Barrel* marked *d* is Turned of a Piece of
Maple or Alder-wood, of fuch a length, that it may
play eafily between the two Wooden *Ribs*; and of
fuch a diameter, that in one revolution of it, fuch a
length of *Girt* may wind about it as fhall be equal
to half the length contained between the fore-end
Iron of the *Tympan*, and the infide of the Rail of the
Inner-Tympan; becaufe two Revolutions of this *Barrel*
muft move the *Carriage* this length of fpace.

This *Barrel* is fitted and faftned upon the Iron *Axis*,
at fuch a diftance from either end, that it may move
round between the Wooden *Ribs* aforefaid.

¶. 17. *Of the* Prefs-Stone.

The *Prefs-Stone* fhould be Marble, though fometimes Mafter *Printers* make fhift with Purbeck, either becaufe they can buy them cheaper, or elfe becaufe they can neither diftinguifh them by their appearance, or know their different worths.

Its thicknefs muft be all the way throughout e-
qual

qual, and ought to be within one half quarter of an Inch the depth of the infide of the *Coffin*; becaufe the matter it is *Bedded* in will raife it high enough. Its length and breadth muſt be about half an Inch lefs than the length and breadth of the infide of the *Cof-fin*: Becaufe *Juſtifiers* of Wood, the length of every fide, and almoſt the depth of the *Stone*, muſt be thruſt between the infides of the *Coffin* and the outfides of the *Stone*, to Wedge it tight and ſteddy in its place, after the *Prefs-man* has *Bedded* it. Its upper-fide, or Face muſt be exactly ſtraight and ſmooth.

I have given you this defcription of the *Prefs-Stone*, becaufe they are thus generally ufed in all *Printing-Houfes*: But I have had ſo much trouble, charge and vexation with the often breaking of *Stones*, either through the carelefnefs or unskilfulnefs (or both) of *Prefs-men*, that neceſſity compell'd me to confider how I might leave them off; and now by long experience I have found, that a piece of *Lignum-vitæ* of the fame fize, and truly wrought, performs the office of a *Stone* in all refpects as well as a *Stone*, and eafes my mind, of the trouble, charge and vexation aforefaid, though the firſt coſt of it be greater.

¶. 18. Of the Plattin *marked* d *in Plate* 9.

The *Plattin* is commonly made of Beechen-Planck, two Inches and an half thick, its length about fourteen Inches, and its breadth about nine Inches. Its fides are Tryed Square, and the Face or under-fide of the *Plattin* Plained exactly ſtraight and ſmooth. Near the four Corners on the upper-fide, it hath

four

four Iron *Hooks* as at *a a a a*, whofe Shanks are
Wormed in.

In the middle of the upper-fide is let in and faftned an Iron Plate called the *Plattin-Plate*, as *b b b b*,
a quarter of an Inch thick, fix Inches long, and four
Inches broad; in the middle of this Plate is made a
fquare Iron Frame about half an Inch high, and half
an Inch broad, as at *c*.   Into this fquare Frame is
fitted the *Stud* of the *Plattin Pan*, fo as it may ftand
fteddy, and yet to be taken out and put in as occafion may require.

The *Stud* marked *d*, is about an Inch thick, and
then fpreads wider and wider to the top (at *e e e e*)
of it, till it becomes about two Inches and an half
wide; and the fides of this fpreading being but about
half a quarter of an Inch thick makes the *Pan*.   In
the middle of the bottom of this *Pan* is a fmall Cen-
ter hole Punch'd for the *Toe* of the *Spindle* to work in.

§. 19. *Of the* Points *and* Point-Screws.

The Points are made of Iron Plates about the
thicknefs of a Queen *Elizabeth* Shilling: It is delineated at e in Plate 9. which is fufficient to fhew the
fhape of it, at the end of this Plate, as at *a*, ftands
upright the Point.   This *Point* is made of a piece of
fmall Wyer about a quarter and half quarter of an
Inch high, and hath its lower end Filed away to a fmall
Shank about twice the length of the thicknefs of the
Plate; fo that a Sholder may remain.   This fmall
Shank is fitted into a fmall Hole made near the end
of the Plate, and Revetted on the other fide, as was
                                                    taught

taught *Numb.* 2. *Fol.* 24. At the other end of the
Plate is filed a long fquare notch in the Plate as at *b c*
quarter and half quarter Inch wide, to receive the
fquare fhank of the *Point-Screws.*

The *Point-Screw* marked f is made of Iron; It
hath a thin Head about an Inch fquare, And a fquare
Shank juft under the Head, an Inch deep, and almoft
quarter and half quarter Inch fquare, that the fquare
Notch in the hinder end of the Plate may flide on
it from end to end of the Notch; Under this fquare
Shank is a round Pin filed with a Male-Screw upon
it, to which is fitted a *Nut* with a Female-Screw in
it, and Ears on its out-fide to twift about, and draw
the Head of the Shank clofe down to the *Tympan,*
and fo hold the *Point-Plate* faft in its Place.

¶. 20. *Of the* Hammer, *defcribed at* h, *and* Sheeps-Foot *defcribed at* i *in Plate* 9.

The *Hammer* is a common *Hammer* about a quarter of a Pound weight; It hath no *Claws* but a *Pen,*
which ftands the *Prefs-man* inftead when the *Chafe*
proves fo big, that he is forced to ufe fmall *Quoins.*

The Figure of the *Sheeps-Foot* is defcription fufficient. Its ufe is to nail and un-nail the *Balls.*

The *Sheeps-Foot* is all made of Iron, with an Ham-
mer-head at one end, to drive the *Ball-Nails* into the
Ball-Stocks, and a Claw at the other end, to draw
the *Ball-Nails* out of the *Ball-Stocks.*

¶. 21. *Of*

6*

¶. 21. *Of the* Foot-ftep, Girts, Stay *of the* Carriage, Stay *of the* Frisket, Ball-Stocks, Paper-Bench, Lye-Trough, Lye-Brufh, Lye-Kettle, Tray *to* wet Paper *in*, Weights to Prefs Paper, Pelts, *or* Leather, Wool *or* Hair, Ball-Nails *or* Pumping-Nails.

The *Foot-Step* is an Inch-Board about a Foot broad, and fixteen Inches long. This Board is nailed upon a piece of Timber about feven or eight Inches high, and is Bevil'd away on its upper-fide, as is alfo the Board on its under-fide at its hither end, that the Board may ftand aflope upon the Floor. It is placed faft on the Floor under the Carriage of the Prefs. Its Office fhall be fhewed when we come to treat of Exercife of the *Prefs-man*.

Girts are Thongs of Leather, cut out of the Back of an Horfe-hide, or a Bulls hide, fometimes an Hogs-hide. They are about an Inch and an half, or an Inch and three quarters broad. Two of them are ufed to carry the *Carriage* out and in. Thefe two have each of them one of their ends nailed to the *Barrel* on the *Spindle* of the *Rounce*, and the other ends nailed to the *Barrel* behind the *Carriage* in the *Planck* of the *Coffin*, and to the *Barrel* on the fore-end of the Frame of the *Coffin*.

The *Stay* of the *Carriage* is fometimes a piece of the fame *Girt* faftned to the outfide of the further *Cheek*,

and

and to the further hinder fide of the Frame of the *Carriage*. It is faftned at fuch a length by the *Prefs man*, that the *Carriage* may ride fo far out, as that the Irons of the *Tympan* may juft rife free and clear off the fore-fide of the *Plattin*.

Another way to ftay the *Carriage* is to let an Iron Pin into the upper-fide of the further Rail of the Frame of the *Ribs*, juft in the place where the further hinder Rail of the *Carriage* ftands projecting over the *Rib-Rail*, when the Iron of the *Tympan* may juft rife free from the Fore-fide of the *Plattin*; for then that projecting will ftop againft the Iron Pin.

The *Stay* of the *Frisket* is made by faftning a Batten upon the middle of the Top-fide of the *Cap*, and by faftning a Batten to the former Batten perpendicularly downwards, juft at fuch a diftance, that the upper-fide of the *Frisket* may ftop againft it when it is turned up juft a little beyond a Perpendicular. When a *Prefs* ftands at a convenient diftance from a Wall, that Wall performs the office of the aforefaid *Stay*.

Ball-Stocks are Turn'd of *Alder* or *Maple*. Their Shape is delineated in Plate 9. at g: They are about feven Inches in Diameter, and have their under fide Turned hollow, to contain the greater quantity of *Wool* or *Hair*, to keep the *Ball-Leathers* plump the longer.

The *Lye-Trough* (delineated in Plate 9. at k) is a Square Trough made of Inch-Boards, about four Inches deep, two Foot four Inches long, and one Foot nine Inches broad, and flat in the Bottom. Its infide is Leaded with Sheet-Lead, which reaches up over the upper Edges of the *Trough*. In the middle of the two ends (for fo I call the fhorteft fides) on the outer fides as *a a*, is faftned a round Iron Pin, which

moves

moves in a round hole made in an Iron Stud with a square Sprig under it, to be drove and faftned into a *Wooden Horfe,*which *Horfe* I need not defcribe, becaufe in Plate aforefaid I have given you the Figure of it.

The *Paper-Bench* is only a common Bench about three Foot eight Inches long, one Foot eight Inches broad, and three Foot four Inches high.

The *Lye Brufh* is made of *Hogs-Briftles* faftned into a Board with Brafs-Wyer, for durance fake: Its Board is commonly about nine Inches long, and four and an half Inches broad; and the length of the Briftles about three Inches.

To perform the Office of a *Lye-Kettle* (which commonly holds about three Gallons) the old-fafhion'd *Chafers* are moft commodious, as well becaufe they are more handy and manageable than *Kettles* with Bails, as alfo becaufe they keep *Lye* longer hot.

The *Tray* to *Wet Paper* in is only a common Butchers Tray, large enough to *Wet* the largeft *Paper* in.

The *Weight* to *Prefs Paper* with, is either Mettal, or Stone, flat on the Bottom, to ly fteddy on the *Paper-Board*: It muft be about 50 or 60 pound weight.

For *Pelts* or *Leather, Ball-Nails* or *Pumping-Nails, Wool* or *Hair, Vellom* or *Parchment* or *Forrel,*the *Prefs-man* generally eafes the *Mafter-Printer* of the trouble of choofing, though not the charge of paying for them: And for *Pafte, Sallad Oyl,* and fuch accidental Requifites as the *Prefs-man* in his work may want, the *Devil* commonly fetches for him.

¶. 22. *Of* Racks *to Hang* Paper *on, and of the* Peel.

Our Mafter-*Printer* muft provide *Racks* to hang *Paper* on to *Dry.* They are made of Deal-board Battens,
                                                    fquare,

fquare, an Inch thick, and an Inch and an half deep, and the length the whole length of the Deal, which is commonly about ten or eleven Foot long, or elfe fo long as the convenience of the Room will allow: The two upper corners of thefe *Rails* are rounded off that they may not mark the *Paper.*

Thefe *Racks* are Hung over Head, either in the *Printing-Houfe,* or *Ware-houfe,* or both, or any other Room that is moft convenient to *Dry Paper* in; they are hung a-thwart two *Rails* an Inch thick, and about three or four Inches deep, which *Rails* are faftned to fome Joyfts or other Timber in the Ceiling by Stiles perpendicular to the Ceiling; Thefe *Rails* ftand fo wide affunder, that each end of the *Racks* may hang beyond them about the diftance of two Foot, and have on their upper edge at ten Inches diftance from one another, fo many fquare Notches cut into them as the whole length of the *Rail* will bear; Into thefe fquare notches the *Racks* are laid parallel to each other with the flat fide downwards, and the Rounded off fide upwards.

The *Peel* is defcribed in Plate 9. at l, which Figure fufficiently fhews what it is; And therefore I fhall need fay no more to it, only its Handle may be longer or fhorter according as the height of the Room it is to be ufed in may require.

¶. 23. *Of* Inck.

The providing of good *Inck,* or rather good *Varnifh* for *Inck,* is none of the leaft incumbent cares upon our *Mafter-Printer,* though Cuftom has almoft made it fo here in *England*; for the procefs of making *Inck* being as well laborious to the Body, as noyfom

fom and ungrateful to the Sence, and by feveral odd accidents dangerous of Firing the Place it is made in, Our *Englifh Mafter-Printers* do generally difcharge themfelves of that trouble; and inftead of having good *Inck*, content themfelves that they pay an *Inck-maker* for good *Inck*, which may yet be better or worfe according to the Confcience of the *Inck-maker*.

That our Neighbours the *Hollanders* who exhibit Patterns of good *Printing* to all the World, are careful and induftrious in all the circumftances of good *Printing*, is very notorious to all Book-men; yet fhould they content themfelves with fuch *Inck* as we do, their Work would appear notwithftanding the other circumftances they obferve, far lefs graceful than it does, as well as ours would appear more beautiful if we ufed fuch *Inck* as they do: for there is many Reafons, confidering how the *Inck* is made with us and with them, why their *Inck* muft needs be better than ours. As *Firft*, They make theirs all of good old *Linfeed-Oyl* alone, and perhaps a little *Rofin* in it fometimes, when as our *Inck-makers* to fave charges mingle many times *Trane-Oyl* among theirs, and a great deal of *Rofin*; which *Trane-Oyl* by its grofsnefs, Furs and Choaks up a *Form*, and by its fatnefs hinders the *Inck* from drying; fo that when the Work comes to the *Binders*, it *Sets off*; and befides is dull, fmeary and unpleafant to the Eye. And the *Rofin* if too great a quantity be put in, and the *Form* be not very *Lean Beaten*, makes the *Inck* turn yellow: And the fame does New *Linfeed-Oyl*.

Secondly, They feldom *Boyl* or *Burn* it to that confiftence the *Hollanders* do, becaufe they not only fave labour and Fewel, but have a greater weight of *Inck*
out

out of the fame quantity of *Oyl* when lefs *Burnt* a-
way than when more *Burnt* away; which want of
Burning makes the *Inck* alfo, though made of good
old *Linfeed Oyl* Fat and Smeary, and hinders its *Drying*;
fo that when it comes to the *Binders* it alfo *Sets off.*

Thirdly, They do not ufe that way of clearing
their *Inck* the *Hollanders* do, or indeed any other way
than meer Burning it, whereby the *Inck* remains
more *Oyly* and *Greafie* than if it were well clarified.

Fourthly, They to fave the *Prefs-man* the labour of
Rubbing the *Blacking* into *Varnifh* on the *Inck-Block,*
Boyl the *Blacking* in the *Varnifh,* or at leaft put the
Blacking in whilft the *Varnifh* is yet *Boyling-hot,* which
fo *Burns* and *Rubifies* the *Blacking,* that it lofes much
of its brisk and vivid black complexion.

Fifthly, Becaufe *Blacking* is dear, and adds little to
the weight of *Inck,* they ftint themfelves to a quantity which they exceed not; fo that fometimes the
Inck proves fo unfufferable *Pale,* that the *Prefs-man* is
forc'd to *Rub* in more *Blacking* upon the *Block*; yet
this he is often fo loth to do, that he will rather ha-
zard the content the Colour fhall give, than take the
pains to amend it: fatisfying himfelf that he can
lay the blame upon the *Inck-maker.*

Having thus hinted at the difference between the
Dutch and *Englifh Inck,* I fhall now give you the Re-
ceipt and manner of making the *Dutch-Varnifh.*

They provide a *Kettle* or a *Caldron,* but a *Caldron*
is more proper, fuch an one as is defcribed in Plate
9. at m. This Veffel fhould hold twice fo much *Oyl*
as they intend to *Boyl,* that the *Scum* may be fome
confiderable time a *Rifing* from the top of the *Oyl* to
the top of the Veffel to prevent danger. This *Cal-*
dron

dron hath a Copper Cover to fit the Mouth of it, and this Cover hath an Handle at the top of it to take it off and put it on by. This *Caldron* is set upon a good strong Iron *Trevet*, and fill'd half full of old *Linseed-Oyl*, the older the better, and hath a good Fire made under it of solid matter, either *Sea Coal*, *Charcoal* or pretty big Chumps of Wood that will burn well without much Flame; for should the Flame rise too high, and the *Oyl* be very hot at the taking off the Cover of the *Caldron*, the fume of the *Oyl* might be apt to take Fire at the Flame, and endanger the loss of the *Oyl* and Firing the House: Thus they let *Oyl* heat in the *Caldron* till they think it is Boyling-hot; which to know, they peel the outer Films of an *Oynion* off it, and prick the *Oynion* fast upon the end of a small long Stick, and so put it into the heating *Oyl*: If it be Boyling-hot, or almost Boyling-hot, the *Oynion* will put the *Oyl* into a Fermentation, so that a Scum will gather on the top of the *Oyl*, and rise by degrees, and that more or less according as it is more or less Hot: But if it be so very Hot that the Scum rises apace, they quickly take the *Oynion* out, and by degrees the Scum will fall.   But if the *Oyl* be Hot enough, and they intend to put any *Rosin* in, the quantity is to every Gallon of *Oyl* half a Pound, or rarely a whole Pound. The *Rosin* they beat small in a *Mortar*, and with an Iron Ladle, or else by an Handful at a time strew it in gently into the *Oyl* lest it make the Scum rise too fast; but every Ladle-full or Handful they put in so leasurely after one another, that the first must be wholly dissolv'd before they put any more in; for else the Scum will Rise too fast, as aforesaid: So that
                                                            you

you may perceive a great care is to keep the Scum down: For if it Boyl over into the Fire never fo little, the whole Body of *Oyl* will take Fire immediately.

If the *Oyl* be Hot enough to *Burn*, they *Burn* it, and that fo often till it be *Hard* enough, which fometimes is fix, feven, eight times, or more.

To *Burn* it they take a long fmall Stick, or double up half a Sheet of Paper, and light one end to fet Fire to the *Oyl*; It will prefently Take if the *Oyl* be Hot enough, if not, they Boyl it longer, till it be.

To try if it be *Hard* enough, they put the end of a Stick into the *Oyl*, which will lick up about three or four drops, which they put upon an Oyfter-fhell, or fome fuch thing, and fet it by to cool, and when it is cold they touch it with their Fore or Middle-Finger and Thumb, and try its confiftence by fticking together of their Finger and Thumb; for if it draw ftiff like ftrong *Turpentine* it is Hard enough, if not, they Boyl it longer, or *Burn* it again till it be fo confolidated.

When it is well Boyled they throw in an Ounce of Letharge of Silver to every four Gallons of *Oyl* to Clarifie it, and Boyl it gently once again, and then take it off the Fire to ftand and cool, and when it is cool enough to put their Hand in, they Strain it through a Linnen Cloath, and with their Hands wring all the *Varnifh* out into a Leaded Stone Pot or Pan, and keeping it covered, fet it by for their ufe; The longer it ftands by the better, becaufe it is lefs fubject to turn Yellow on the Paper that is Printed with it.

This is the *Dutch* way of making *Varnifh*, and the way the Englifh *Inck-makers* ought to ufe.

Note, Firft, That the *Varnifh* may be made without *Burning* the *Oyl*, *viz.* only with well and long
Boyl-

Boyling it; for *Burning* is but a violent way of Boyling, to confolidate it the fooner.

Secondly, That an *Apple* or a *Cruft* of *Bread, &c.* ftuck upon the end of a Stick inftead of an *Oynion* will alfo make the Scum of the *Oyl* rife: For it is only the Air contained in the Pores of the *Apple, Cruft* or *Oynion,* &c. preffed or forced out by the violent heat of the *Oyl,* that raifes the many Bubbles on the top of the *Oyl:* And the connection of thofe Bubbles are vulgarly called *Scum.*

Thirdly, The Englifh *Inck-makers* that often make *Inck,* and that in great quantities, becaufe one Man may ferve all *England,* inftead of fetting a *Caldron* on a *Trevet,* build a *Furnace* under a great *Caldron,* and Trim it about fo with Brick, that it Boyls far fooner and more fecurely than on a *Trevet*; becaufe if the *Oyl* fhould chance to Boyl over, yet can it not run into the Fire, being Fenced round about with Brick as a-forefaid, and the *Stoking-hole* lying far under the *Caldron.*

Fourthly, When for want of a *Caldron* the *Mafter-Printer* makes *Varnifh* in a *Kettle,* He provides a great piece of thick *Canvafs,* big enough when three or four double to cover the *Kettle,* and alfo to hang half round the fides of the *Kettle*: This *Canvafs* (to make it more foluble) is wet in Water, and the Water well wrung out again, fo that the *Canvafs* remains only moift: Its ufe is to throw flat over the Mouth of the *Kettle* when the *Oyl* is *Burning,* to keep the fmoak in, that it may ftifle the Flame when they fee caufe to put it out. But the Water as was faid before, muft be very well wrung out of the *Canvafs,* for fhould but a drop or two fall from the fides of it into the *Oyl* when it is Burning, it will fo enrage the *Oyl,* and raife the Scum, that it might endanger the working over the top of the *Kettle.*

Having fhewn you the *Mafter-Printers* Office, I account it fuitable to proper Method, to let you know how the *Letter-Founder* Cuts the *Punches,* how the *Molds* are made, the *Matrices* Sunck, and the *Letter Caft* and *Dreft,* for all thefe Operations precede the *Compofiters* Trade, as the *Compofiters* does the *Prefs-mans*; wherefore the next *Exercifes* fhall be (God willing) upon *Cutting* of the *Steel-Punches.* *ME-*

MECHANICK EXERCISES:

Or, the Doctrine of

𝕳𝖆𝖓𝖉𝖞-𝖜𝖔𝖗𝖐𝖘.

Applied to the Art of

𝕷𝖊𝖙𝖙𝖊𝖗-𝕮𝖚𝖙𝖙𝖎𝖓𝖌.

PREFACE.

L Etter-Cutting *is a Handy-Work hitherto kept so conceal'd among the Artificers of it, that I cannot learn any one hath taught it any other ; But every one that has used it, Learnt it of his own Genuine Inclination. Therefore, though I cannot (as in other Trades) describe the general Practice of Work-men, yet the Rules I follow I shall shew here, and have as good an Opinion of these Rules, as those have that are shyest of discovering theirs. For, indeed, by the appearance of some Work done, a judicious Eye may doubt whether they go by any Rule at all, though Geometrick Rules, in no Practice whatever, ought to be more nicely or exactly observed than in this.*

§ 12.

§. 12. ¶. 1. *Of Letter-Cutters Tools.*

THe making of *Steel Punches* is a Branch of
the *Smith*'s Trade: For, as I told you in the
Preface to *Numb.* 1. The *Black-Smith*'s Trade com-
prehends all Trades that ufe either Forge or File,
from the *Anchor-Smith,* to the *Watch-maker*: They
all working by the fame Rules, though not with
equal exactnefs; and all ufing the fame Tools,
though of different Sizes from thofe the Common
Black-Smith ufes; and that according to the various
purpofes they are applied, &c. Therefore, indeed,
a *Letter-Cutter* fhould have a Forge fet up, as by
Numb. 1. But fome *Letter-Cutters* may feem to fcorn
to ufe a Forge, as accounting it too hard Labour,
and Ungenteel for themfelves to officiate at. Yet
they all well know, that though they may have a
common *Black-Smith* perform their much and heavy Work, that many times a Forge of their own at
Hand would be very commodious for them in feveral accidental little and light Jobs, which (in a Train
of Work) they muft meet withal.

But if our *Letter-Cutter* will have no Forge, yet
he muft of neceffity accommodate himfelf with a
Vice, Hand-Vice, Hammers, Files, Small and *Fine
Files* (commonly called *Watch-makers Files*) of thefe
he faves all, as they wear out, to fmooth and bur-
nifh the Sides and Face of his Letter with, as fhall
be fhewed; *Gravers,* and *Sculpters* of all forts, an
Anvil, or a *Stake,* an *Oyl-ftone,* &c. And of thefe,
fuch as are fuitable and fizable to the feveral Letters
he

Plate 10

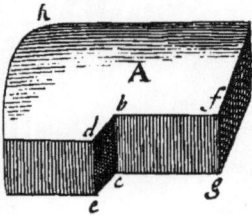

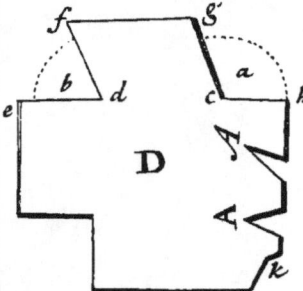

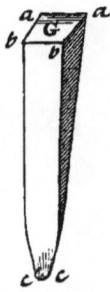

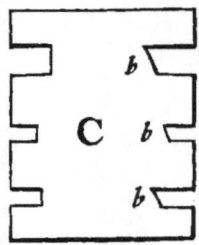

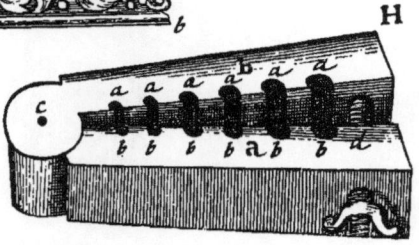

he is to Cut.   Thefe, or many of thefe Tools, being
defcribed in *Numb.* 1. I refer my Reader thither,
and proceed to give an account of fome Tools peculiar to the *Letter-Cutter*, though not of particular
ufe to the Common *Black-Smith*.

¶. 2. *Of the* Ufing-File.

This *File* is about nine or ten Inches long, and
three or four Inches broad, and three quarters of
an Inch thick: The two broad fides muft be exactly flat and ftraight: And the one fide is commonly
cut with a *Baftard-Cut*, the other with a *Fine* or
Smooth Cut. (See *Numb.* 1. *Fol.* 14, 15.) Its ufe is
to *Rub* a piece of Steel, Iron, or Brafs, &c. flat and
ftraight upon, as fhall be fhewed hereafter.

In chufing it, you muft fee it be exactly Flat and
Straight all its Length and Breadth: For if it in any
part Belly out, or be Hollow inwards, what is Rub-
bed upon it will be Hollow, Rubbing on the Bellying part; and Bellying, Rubbing on the Hollow
part.   You muft alfo fee that it be very Hard; and
therefore the thickeft *Ufing-Files* are likelieft to
prove beft, becaufe the thin commonly Warp in
Hardning.

¶. 3. *Of the* Flat-Gage.

The *Flat-Gage* is defcribed in *Plate* 10. at A. It
is made of a flat piece of Box, or other Hard Wood.
Its Length is three Inches and an half, its Breadth two
Inches and an half, ànd its Thicknefs one Inch and
                                                          an

an half. This is on the Flat, firft made Square, but afterwards hath one of its Corners (as *h*) a little rounded off, that it may the eafier comply with the Ball of the Hand. Out of one of its longeft Sides, *viz.* that not rounded off, is Cut through the thicknefs of it an exact Square, whofe one fide *b f*, *c g* is about an Inch and three quarters long; and its other fide *b d*, *c e* about half an Inch long. The Depth of thefe Sides and their Angle is exactly Square to the top and bottom of the upper and under Superficies of the *Flat-Gage*.

Its Ufe is to hold a Rod of Steel, or Body of a *Mold*, *&c.* exactly perpendicular to the Flat of the *Ufing-File*, that the end of it may rub upon the *Ufing-File*, and be Filed away exactly Square, and that to the Shank; as fhall more at large be fhewed in §. 2. ¶. 3.

¶. 4. *Of the* Sliding Gage.

The *Sliding Gage* is defcribed in *Plate* 10. at *Fig.* B. It is a Tool commonly ufed by *Mathematical Inftrument-Makers*, and I have found it of great ufe in *Letter-Cutting*, and making of *Molds*, *&c. a a* the Beam, *b* the Tooth, *c c* the Sliding Socket, *d d d d* the Shoulder of the Socket.

Its Ufe is to meafure and fet off Diftances between the Sholder and the Tooth, and to mark it off from the end, or elfe from the edge of your Work.

I always ufe two or three of thefe *Gages*, that I need not remove the Sholder when it is fet to a Diftance which I may have after-ufe for; as fhall in Working be fhewed more fully.

¶. 5.

¶. 5. *Of the* Face-Gages, *marked* C *in Plate* 10.

The *Face-Gage* is a Square Notch cut with a File
into the edge of a thin Plate of Steel, Iron, or Brafs,
the thicknefs of a piece of common Latton, and the
Notch about an *Englifh* deep.　There be three of
thefe Gages made, for the Letters to be cut on one
Body; but they may be all made upon one thin Plate,
the readier to be found, as at D. As firft, for the Long
Letters; Secondly, for the Affending Letters; And
Thirdly, for the Short Letters. The Length of thefe
feveral Notches, or Gages, have their Proportions
to the Body they are cut to, and are as follows. We
fhall imagine (for in Practice it cannot well be perform'd, unlefs in very large Bodies) that the Length
of the whole Body is divided into forty and two
equal Parts.

The *Gage* for the Long Letters are the length of
the whole Body, *viz.* forty and two equal Parts.
The *Gage* for the Affending Letters, *Roman* and *Italica*, are five Seventh Parts of the Body, *viz.* thirty
Parts of Forty two, and thirty and three Parts for
Englifh Face.　The *Gage* for the Short Letters are
three Seventh Parts of the whole Body, *viz.* eighteen
Parts of Forty two for the *Roman* and *Italica*, and
twenty two Parts for the *Englifh* Face.

It may indeed be thought impoffible to divide a
Body into feven equal Parts, and much more difficult to divide each of thofe feven equal Parts into
fix equal Parts, which are Forty two, as aforefaid,
efpecially if the Body be but fmall; but yet it is
poffible

7*

poffible with curious Working: For feven thin Spaces may be Caft and Rubb'd to do it. And for dividing each of the thin Spaces into fix equal Parts, you may Caft and Rub Full Point . to be of the thicknefs of one thin Space, and one fixth part of a thin Space: And you may Caft and Rub : to be the thicknefs of one thin Space, and two fixth parts of a thin Space: And you may Caft and Rub , to be the thicknefs of one thin Space, and three fixth parts of a thin Space: And you may Caft and Rub - to be the thicknefs of one thin Space, and four fixth parts of a thin Space: And you may Caft and Rub ; to be the thicknefs of one thin Space, and five fixth parts of a thin Space.

The reafon why I propofe . to be Caft and Rubb'd one fixth part thicker than a thin Space, is only that it may be readily diftinguifhed from : , - ; which are two fixth parts, three fixth parts, four fixth parts, five fixth parts thicker than a thin Space. And for fix fixth parts thicker than a thin Space, two thin Spaces does it.

The manner of adjufting thefe feveral Sixth Parts of Thickneffes is as follows. You may try if fix . exactly agree, and be even with feven thin Spaces; (or, which is all one, a Body) for then is each of thofe fix . one fixth part thicker than a thin Space, becaufe it drives out a thin Space in fix thin Spaces. And you may try if fix : be equal to a Body and one thin Space; for then is each : two fixth parts thicker than a thin Space. If fix , be equal to nine thin Spaces, then each , is three fixth parts of a thin Space thicker than a thin Space. If fix -

be

be equal to ten thin Spaces, then each - is four fixth parts of a thin Space thicker than a thin Space. If fix ; be equal to eleven thin Spaces, then each ; is five fixth parts of a thin Space thicker than a thin Space.

Now, as aforefaid, a thin Space being one feventh part of the Body, and the thin Space thus divided, you have the whole Body actually divided into forty and two equal parts, as I have divided them in my Drafts of Letters down the Sides, and in the Bottom Line.

Though I have thus fhewed how to divide a thin Space into fix equal Parts, yet when the Letter to be Cut proves of a fmall Body, the thin Space divided into two equal Parts may ferve: If it prove bigger, into three or four equal Parts: And of the largeft Bodies, they may be divided into fix, as aforefaid.

If now you would make a *Gage* for any number of thin Spaces and Sixth Parts of a thin Space, you muft take one thin Space lefs than the number of thin Spaces propofed, and add . : , - ; according as the number of fixth Parts of a thin Space require; and to thofe complicated Thickneffes you may file a fquare Notch on the edge of the thin Plate aforefaid, which fhall be a ftanding Gage or Meafure for that number of thin Spaces and fixth Parts of a thin Space.

All the Exception againft this way of Meafuring is, that thin Spaces caft in Metal may be fubject to bow, and fo their Thickneffes may prove deceitful. But, in Anfwer to that, I fay, you may, if you will,

Caft

Caft I for two thin Spaces thick, e for three
thin Spaces thick, S for four thin Spaces thick,
L for five thin Spaces thick, D for fix thin Spaces thick, or any other Letters near thefe feveral
Thickneffes, as you think fit; only remember, or
rather, make a Table of the number of thin Spaces
that each Letter on the Shank is Caft for. And by
complicating the Letters and Points, as aforefaid,
you will have any Thicknefs, either to make a Gage
by, or to ufe otherwife.

On the other Edge of the *Face-Gage* you may file
three other Notches, of the fame Width with thofe
on the former Edge, for the Long, the Affending,
and Short Letters. But though the two fides of
each of thefe Notches are parallel to each other, yet
is not the third fide fquare to them, but hath the
fame Slope the *Italick* hath from the *Roman* ; as you
may fee in the Figure at *b b b.*

¶. 6. *Of* Italick, *and other* Standing Gages. ·

Thefe *Gages* are to meafure (as aforefaid) the
Slope of the *Italick* Stems, by applying the Top and
Bottom of the *Gage* to the Top and Bottom Lines
of the Letters, and the other Side of the *Gage* to
the Stem: for when the Letter complies with thefe
three fides of the *Gage* that Letter hath its true
Slope.

The manner of making thefe *Gages* (and indeed
all other *Angular Gages*) is thus.

Place one Point of a Pair of Steel *Dividers* upon
the thin Plate aforefaid, at the Point *c* or *d* (in
Fig.

Fig. D in *Plate* 10.) and with the other Point de-
fcribe a fmall fine Arch of a Circle; as, *e f* or *g h*.
In this Arch of the Circle muft be fet off on the *Gage
a* 110 Degrees, and on the *Gage b* 70 Degrees,
and draw from the Centres *c* and *d* two ftraight
Lines through thofe numbers of Degrees: Then
Filing away the Plate between the two Lines, the
Gages are finifhed.

To find the Meafure of this, or any other number of Degrees, do thus; Defcribe a Circle on a
piece of Plate-Brafs of any Radius (but the larger
the better) draw a ftraight Line exactly through
the Centre of this Circle, and another ftraight Line
to cut this ftraight Line at right Angles in the Centre,
through the Circle; fo fhall the Circle be divided
into four Quadrants: Then fix one Foot of your
Compaffes (being yet unftirr'd) in one of the Points
where any of the ftraight Lines cuts the Circle, and
extend the moving Foot of your Compaffes where
it will fall in the Circle, and make there a Mark,
which is 60 Degrees from the fixed Foot of the
Compaffes: Then fix again one Foot of your Compaffes in the Interfection of the ftraight Line and
Circle that is next the Mark that was made before,
and extend the moving Foot in the fame Quadrant
towards the ftraight Line where you firft pitch'd the
Foot of your Compaffes, and with the moving Foot
make another Mark in the Circle. Thefe two Marks
divide the Quadrant into three equal Parts: The fame
way you may divide the other three Quadrants; fo
fhall the whole Circle be divided into twelve equal
Parts; and each of thefe twelve equal parts contain

tain an Arch of thirty Degrees: Then with your
Dividers divide each of thefe 30 Degrees into three
equal Parts, and each of thefe three equal Parts into
two equal Parts, and each of thefe two equal Parts
into five equal Parts, fo fhall the Circle be divided
into 360 equal Parts, for your ufe.

To ufe it, defcribe on the Centre of the Circle an
Arch of almoft a Semi-Circle: This Arch muft be
exactly of the fame Radius with that I prefcribed to
be made on the *Gages a b*, from *e* to *f*, and from *g*
to *h*; then count in your Circle of Degrees from
any Diametral Line 110 Degrees; and laying a
ftraight Ruler on the Centre, and on the 110 Degrees aforefaid, make a fmall Mark through the
fmall Arch; and placing one Foot of your Compaffes at the Interfection of the fmall Arch, with the
Diametral Line, open the other Foot to the Mark
made on the fmall Arch for 110 Degrees, and tranf-
fer that Diftance to the fmall Arch made on the
Gage: Then through the Marks that the two Points
of your Compaffes make in the fmall Arch on the
Gage, draw two ftraight Lines from the Centre *c:*
and the Brafs between thofe two ftraight Lines being filed away, that *Gage* is made. In like manner
you may fet off any other number of Degrees, for
the making of any other *Gage*.

In like manner, you may meafure any Angle in
the Drafts of Letters, by defcribing a fmall Arch on
the Angular Point, and an Arch of the fame Radius
on the Centre of your divided Circle: For then,
placing one Foot of your Compaffes at the Interfection of the fmall Arch with either of the ftraight
Lines

Lines proceeding from the Angle in the Draft, and extending the other Foot to the Interſection of the ſmall Arch, with the other ſtraight Line that proceeds from the Angle, you have between the Feet of your Compaſſes, the Width of the Angle; and by placing one Foot of your Compaſſes at the Interſection of any of the ſtraight Lines that proceed from the Centre of the divided Circle, and the ſmall Arch you made on it, and making a Mark where the other Foot of your Compaſſes falls in the ſaid ſmall Arch, you may, by a ſtraight Ruler laid on the Centre of the divided Circle, and the Mark on the ſmall Arch, ſee in the Limb of the Circle the number of Degrees contained between the Diametral, or ſtraight Line and the Mark.

If you have already a dividing-Plate of 360 Degrees, of a larger Radius than the Arch on your *Gage*, you may ſave your ſelf the labour of dividing a Circle (as aforeſaid,) and work by your dividing-Plate as you were directed to do with the Circle that I ſhewed you to divide.

In theſe Documents I have expoſed my ſelf to a double Cenſure; Firſt, of *Geometricians:* Secondly, of *Letter-Cutters. Geometricians* will cenſure me for writing anew that which almoſt every young Beginner knows: And *Letter-Cutters* will cenſure me for propoſing a Rule for that which they dare pretend they can do without Rule.

To the *Geometricians* I croſs the Cudgels: yet I writ this not to them; and I doubt I have written ſuperfluouſly to *Letter-Cutters*, becauſe I think few of them either will or care to take pains to underſtand
                                                    theſe

thefe fmall Rudiments of *Geometry.* If they do, and
be ingenious, they will thank me for difcovering this
Help in their own Way, which few of them know.
For by this Rule they will not only make Letters
truer, but alfo quicker, and with lefs care; becaufe
they fhall never need to ftamp their *Counter-Punch*
in Lead, to fee how it pleafes them; which they
do many times, before they like their *Counter-Punch*,
(be it of A *A* V v W w *V W,* and feveral other Letters) and at laft finifh their *Counter-Punch* but with
a good Opinion they have that it may do well,
though they frequently fee it does not in many Angular Letters on different Bodies Cut by the fame
Hand. And were *Letter-Cutting* brought to fo common Practice as *Joynery, Cabinet-making,* or *Mathematical Inftrument-making,* every young Beginner
fhould then be taught by Rules, as they of thefe
Trades are; becaufe *Letter-Cutting* depends as much
upon Rule and Compafs as any other Trade does.

You may in other places, where you find moft
Convenience (as at *i*) make a Square, which may
ftand you in ftead for the Squaring the Face and
Stems of the *Punch* in *Roman* Letters, and alfo in
many other Ufes.

And you may make *Gages,* as you were taught
before to try the *Counter-Punches* of Angular Letters;
as, A K M N V X Y Z, *Romans* and *Italicks, Capitals* and *Lower-Cafe.* But then, that you may know
each diftinct *Gage,* you may engrave on the feveral
refpective *Gages,* at the Angle, A *A* 4 *&c.* For by
examining by the Drafts of Letters, what Angle
their Infides make, you may fet that Angle off, and
make

make the *Gage* as you were taught before, in the *Gage* for the Slope of *Italicks*.

¶. 7. *Of the* Liner.

The *Liner* is marked E in *Plate* 10. It is a thin Plate of Iron or Brafs, whofe Draft is fufficient to exprefs the Shape. The Ufe of it is on the under-edge *a b* (which is about three Inches long) and is made truly ftraight, and pretty fharp or fine; that being applied to the Face of a *Punch*, or other piece of Work, it may fhew whether it be ftraight or no.

¶. 8. *Of the* Flat-Table.

The *Flat-Table* at F in *Plate* 10. The Figure is there fufficient. All its Ufe is the Table F, for that is about one Inch and an half fquare, and on its Su-perficies exactly ftraight and flat. It is made of Iron or Brafs, but Brafs moft proper. Its Ufe is to try if the Shank of a *Punch* be exactly Perpendicu-lar to its Face, when the Face is fet upon the *Table*; for if the Shank ftand then directly upright to the Face of the *Table*, and lean not to any fide of it, it is concluded to be perpendicular.

It hath feveral other Ufes, which, when we come to *Cafting* of *Letters*, and *Juftifying* of *Matrices*, fhall be fhewn.

¶. 9.

¶. 9. *Of the* Tach.

The *Tach* is a piece of Hard-Wood, (Box is very good) about three Inches broad, fix Inches long, and three quarters of an Inch thick. About half its Length is faftned firm down upon the *Work-Bench*, and its other half projects over the hither Edge of it. It hath three or four Angular Notches on its Fore-end to reft and hold the Shank of a *Punch* ftea-dy when the End of the *Punch* is fcrewed in the *Hand-Vice*, and the *Hand-Vice* held in the Left Hand, while the *Work-man* Files or Graves on it with his Right Hand.

Inftead of Faftning the *Tach* to the *Bench*, I *Saw* a fquare piece out of the further half of the *Tach*, that it may not be too wide for the Chaps of the *Vice* to take and fcrew that narrow End into the Chaps of the *Vice*, becaufe it fhould be lefs cumberfome to my *Work-Bench*.

¶. 10. *Of Furnifhing the* Work-Bench.

The *Work-man* hath all his great *Files* placed in Leather Noofes, with their Handles upwards, that he may readily diftinguifh the *File* he wants from another *File*. Thefe Noofes are nailed on a Board that Cafes the Wall on his Right Hand, and as near his *Vice* as Convenience will admit, that he may the readier take any *File* he wants.

He hath alfo on his Right Hand a Tin Pot, of about a Pint, with fmall *Files* ftanding in it, with their

their Handles downwards, that their Blades may be the readier feen. Thefe fmall *Files* are called *Watch-makers Files*, and the *Letter-Cutter* hath occafion to ufe thefe of all Shapes, *viz. Flat, Pillar, Square, Triangu-lar, Round, Half-Round, Knife-Files, &c.*

He alfo provides a fhallow fquare Box, of about five Inches long, and three Inches broad, to lay his fmall Inftruments in; as, his *Gages*, his *Liner*, fome common *Punches, &c.* This Box he places before him, at the further fide of the *Work-Bench*.

He alfo provides a good *Oyl-Stone*, to fharpen his *Gravers* and *Sculpters* on. This he places at fome diftance from the *Vice*, on his Left Hand.

§. 13. ¶. 1. *Of* Letter-Cutting.

The *Letter-Cutter* does either Forge his *Steel-Punches*, or procures them to be forged; as I fhewed, *Numb.* 1. *Fol.* 8, 9, 10. in *Vol.* I. *&c.* But great care muft be taken, that the Steel be found, and free from Veins of Iron, Cracks and Flaws, which may be difcerned; as I fhewed in *Numb.* 3. *Vol.* I. For if there be any Veins of Iron in the Steel, when the Letter is Cut and Temper'd, and you would Sink the *Punch* into the Copper, it will batter there: Or it will Crack or Break if there be Flaws.

If there be Iron in it, it muft with the Chiffel be fplit upon a good Blood-Red-Heat in that place, and the Iron taken or wrought out; and then with another, or more Welding Heat, or Heats, well doubled up, and laboured together, till the Steel become a found entire piece. This Operation *Smiths* call *Well Currying of the Steel.* If

If there be Flaws in it, you muſt alſo take good Welding Heats, ſo hot, that the contiguous ſides of the Flaws may almoſt Run: for then, ſnatching it quickly out of the Fire, you may labour it together till it become cloſe and ſound.

Mr. *Robinſon*, a *Black-Smith* of *Oxford*, told me a way he uſes that is ingenious, and ſeems rational: For if he doubts the Steel may have ſome ſmall Flaws' that he can ſcarce diſcern, he takes a good high Blood-Red Heat of it, and then twiſts the Rod or Bar (as I ſhewed, *Numb.* 3. *Vol.* I.) which Twiſting winds the Flaws about the Body of the Rod, and being thus equally diſpoſed, more or leſs, into the Out-ſides of the Rod, according as the Poſition of the Flaw may be, allows an equal Heat on all ſides to be taken, becauſe the Out-ſides heat faſter than the In-ſide; and therefore the Out-ſides of the Steel are not thus ſo ſubject to Burn, or Run, as if it ſhould be kept in the Fire till the Middle, or In-ſide of it ſhould be ready to Run. And when the Steel is thus well welded, and ſoundly laboured and wrought together with proper Heats, he afterwards reduces it to Form.

Now, that I may be the better underſtood by my Reader as he reads further, I have, in *Plate* 10. at *Fig.* G deſcribed the ſeveral Parts of the *Punch*; which I here explain.

G The Face.

a a, b b The Thickneſs.

a b, a b The Heighth.

a c, b c, b c The Length of the Shank, about an Inch and three quarters long.

c c c The Hammer-End.   This

This is no ſtrict Length for the Shank, but a convenient Length; for ſhould the Letter Cut on the
Face be ſmall, and conſequently, the Shank ſo too,
and the Shank much longer, and it (as ſeldom it is)
not Temper'd in the middle, it might, with Punching into Copper, bow in the middle, either with the
weight of the Hammer, or with light reiterated
Blows: And ſhould it be much ſhorter, there might
perhaps Finger-room be wanting to manage and
command it while it is Punching into the Copper.
But this Length is long enough for the biggeſt Letters, and ſhort enough for the ſmalleſt Letters.

The Heighth and Thickneſs cannot be aſſign'd in
general, becauſe of the diverſity of Bodies, and
Thickneſs of Letters: Beſides, ſome Letters muſt be
Cut on a broad Face of Steel, though, when it is
Cut, it is of the ſame Body; as all Letters are,
to which *Counter-Punches* are uſed; becauſe the
Striking the *Counter-Punch* into the Face of the
Punch will, if it have not ſtrength enough to contain it, break or crack one or more ſides of the
Punch, and ſo ſpoil it. But if the Letter be wholly
to be Cut, and not Counter-Punch'd, as I ſhall here-
after hint in general what Letters are not, then the
Face of the *Punch* need be no bigger, or, at leaſt,
but a ſmall matter bigger than the Letter that is to
be cut upon it.

Now, If the Letter be to be Counter-punch'd,
the Face of the *Punch* ought to be about twice the
Heighth, and twice the Thickneſs of the Face of the
Counter-Punch; that ſo, when the *Counter-Punch* is
ſtruck juſt on the middle of the Face of the *Punch*, a
con-

8

convenient Subftance, and confequently, Strength
of Steel on all its Sides may be contained to refift
the Delitation, that the Sholder or Beard of the
Counter-Punch finking into it, would elfe make.

If the *Letter-Cutter* be to Cut a whole Set of
Punches of the fame Body of *Roman* and *Italica*, he
provides about 240 or 260 of thefe *Punches*, becaufe fo many will be ufed in the *Roman* and *Italica
Capitals* and *Lower-Cafe, Double-Letters, Swafh-Letters,
Accented Letters, Figures, Points, &c.* But this number of *Punches* are to have feveral Heighths and
Thickneffes, though the Letters to be Cut on them
are all of the fame Body.

What Heighth and Thicknefs is, I have fhewed
before in this §, but not what Body is; therefore I
fhall here explain it.

By Body is meant, in *Letter-Cutters, Founders* and
Printers Language, the Side of the Space contained
between the Top and Bottom Line of a Long Letter.
As in the Draft of Letters, the divided Line on the
Left Hand of A is divided into forty and two equal
Parts; and that Length is the Body, thus: J being
an Afcending and Defcending Letter, *viz.* a long
Letter, ftands upon forty two Parts, and therefore
fills the whole Body.

There is in common Ufe here in *England,* about
eleven Bodies, as I fhewed in §. 2. ¶. 2. of this Volumne.

I told you even now, that all the *Punches* for the
fame Body muft not have the fame Heighth and
Thicknefs: For fome are Long; as, J j Q, and feveral others; as you may fee in the Drafts of Letters: and thefe long Letters ftand upon the whole
Heighth of the Body.   The

The Afcending and Defcending Letters reach from the Foot-Line, up to the Top-Line; as all the Capital Letters are Afcending Letters, and fo are many of the Lower-Cafe Letters; as, b d f, and feveral others. The Defcending Letters are of the fame Length with the Afcending Letters; as, g p q and feveral others. Thefe are contained between the Head-Line and the Bottom-Line. The Short Letters are contained between the Head-Line and the Bottom-line. Thefe are three different Sizes of Heighth the *Punches* are made to, for Letters of the fame Body. But in proper place I fhall handle this Subject more large and diftinctly.

And as there is three Heighths or Sizes to be confidered in Letters Cut to the fame Body, fo is there three Sizes to be confidered, with refpect to the Thickneffes of all thefe Letters, when the *Punches* are to be Forged: For fome are m thick; by m thick is meant m *Quadrat* thick, which is juft fo thick as the Body is high: Some are n thick; that is to fay, n *Quadrat* thick, *viz.* half fo thick as the Body is high: And fome are *Space* thick; that is, one quarter fo thick as the Body is high; though Spaces are feldom Caft fo thick, as fhall be fhewed when we come to *Cafting:* and therefore, for diftinction fake, we fhall call thefe Spaces, Thick Spaces.

The firft three Sizes fit exactly in Heighth to all the Letters of the fame Body; but the laft three Sizes fit not exactly in Thicknefs to the Letters of the fame Body; for that fome few among the Capitals are more than m thick, fome lefs than m thick, and more than n thick; and fome lefs than n thick, and

more

more than Space thick; yet for Forging the *Punches*, thefe three Sizes are only in general Confidered, with Exception had to Æ Æ ℺, and moft of the Swafh Letters; which being too thick to ftand on an m, muft be Forged thicker, according to the Work-man's Reafon.

After the Work-man has accounted the exact number of Letters he is to Cut for one Set, he confiders what number he fhall ufe of each of thefe feveral Sizes in the *Roman,* and of each of thefe feveral Sizes in the *Italick*; (for the *Punches* of *Romans* and *Italicks,* if the Body is large, are not to be Forged to the fame fhape, as fhall be fhewed by and by) and makes of a piece of Wood one Pattern of the feveral Sizes that he muft have each number Forged to. Upon every one of thefe Wooden Patterns I ufe to write with a Pen and Ink the number of *Punches* to be Forged of that Size, left afterwards I might be troubled with Recollections.

I fay (for Example) He confiders how many long Letters are m thick, how many long Letters are n thick, and how many long Letters are Space thick, in the *Roman*; and alfo confiders which of thefe muft be Counter-punch'd, and which not: For (as was faid before) thofe Letters that are to be Counter-punch'd are to have about twice the Heighth and twice the Thicknefs of the Face of the *Counter-Punch,* for the Reafon aforefaid. But the Letters not to be Counter-punch'd need no more Subftance but what will juft contain the Face of the Letter; and makes of thefe three Sizes three Wooden Patterns, of the exact Length, Heighth and Thicknefs that the Steel *Punches* are to be Forged to.   He

He alfo counts how many are Afcendents and
Defcendents, m-thick, n-thick, and Space thick;
ftill confidering how many of them are to be Counter-punch'd, and how many not; and makes Wooden
Patterns for them.

The like he does for fhort Letters; and makes
Wooden Patterns for them, for Steel *Punches* to be
Forged by.

And as he has made his Patterns for the *Roman*,
fo he makes Patterns for the *Italick* Letters alfo;
for the fame fhap'd *Punches* will not ferve for *Italick*, unlefs he fhould create a great deal more Work
to himfelf than he need do: For *Italick Punches* are
not all to be Forged with their fides fquare to one
another, as the *Romans* are; but only the higheft and
loweft fides muft ftand in Line with the higheft and
loweft fides of the *Roman*; but the Right and Left
Hand fides ftand not parallel to the Stems of the *Roman*, but muft make an Angle of 20 Degrees with
the *Roman* Stems: fo that the Figure of the Face of
the *Punch* will become a *Rhomboides*, as it is called
by *Geometricians*, and the Figure of this Face is the
Slope that the *Italick* Letters have from the *Roman*,
as in proper place fhall be further fhewed.   Now,
fhould the *Punches* for thefe Letters be Forged with
each fide fquare to one another, the *Letter-Cutter*
would be forced to fpend a great deal of Time, and
take great pains to File away the fuperfluous Steel
about the Face of the Letter when he comes to the
Finifhing of it, efpecially in great Bodied Letters.
Yet are not all the *Italick* Letters to be Forged on
the Slope; for the *Punches* of fome of them, as the

m n,

8*

m n, and many others, may have all, or, at leaft,
three of their fides, fquare to one another, though
their Stems have the common Slope, becaufe the
ends of their Beaks and Tails lie in the fame, perpendicular with the Outer Points of the Bottom and
Top of their Stems, as is fhewed in the Drafts of
Letters.

Though I have treated thus much on the Forging
of Punches, yet muft all what I have faid be under-
ftood only for great Bodied *Punches*; *viz.* from
the *Great Primer,* and upwards. But for fmaller Bodies; as *Englifh,* and downwards, the *Letter-Cutter*
generally, both for *Romans* and *Italicks,* gets fo many fquare Rods of Steel, Forged out of about two
or three Foot in Length, as may ferve his purpofe;
which Rods he elects as near his Body and Sizes as
his Judgment will ferve him to do; and with the
edge of a Half-round File, or a Cold-Chiffel, cuts
them into fo many Lengths as he wants *Punches.*
Nay, many of thefe Rods may ferve for fome of the
fmall Letters in fome of the greater Bodies; and alfo, for many of their *Counter-Punches.*

Having thus prepared your *Punches,* you muft
Neal them, as I fhewed in *Numb.* 3. *Vol.* I.

¶. 2. *Of* Counter-Punches.

The *Counter-Punches* for great Letters are to be
Forged as the *Letter-Punches*; but for the fmaller
Letters, they may be cut out of Rods of Steel, as
aforefaid. They muft alfo be well Neal'd, as the
Punches. Then muft one of the ends be Filed away
on

on the out-fide the Shank, to the exact fhape of the in-fide of the Letter you intend to Cut.   For Example, If it be *A* you would Cut; This *Counter-Punch* is eafie to make, becaufe it is a Triangle; and by meafuring the In-fide of the Angle of *A* in the Draft of Letters, as you were taught, §. 12. ¶. 6. you may make on your Standing *Gage-Plate* a *Gage* for that Angle: So that, let the Letter to be Cut be of what Body you will, from the leaft, to the biggeft Body, you have a Standing *Gage* for this *Counter-Punch*, fo oft as you may have occafion to Cut A.

The *Counter-Punch* of *A* ought to be Forged Triangularly, efpecially towards the Punching End, and Tryed by the *A Gage*, as you were taught to ufe the Square, *Numb.* 3. *Vol.* I.   Yet, for this and other Triangular *Punches*, I commonly referve my worn out three fquare Files, and make my *Counter-Punch* of a piece of one of them that beft fits the Body I am to Cut.

Having by your *A-Gage* fitted the Top-Angle and the Sides of this *Counter-Punch*, you muft adjuft its Heighth by one of the three *Face-Gages* mentioned in §. 12. ¶. 5. *viz.* by the Afcending *Face-Gage*; for *A* is an Afcending Letter.   By Adjufting, I do not mean, you muft make the *Counter-Punch* fo high, as the Depth of the Afcending *Face-Gage*; becaufe in this Letter here is to be confidered the Top and the Footing, which ftrictly, as by the large Draft of *A*, make both together five fixth Parts of a thin Space: Therefore five fixth Parts muft be abated in the Heighth of your *Counter-Punch*, and it muft be but four thin Spaces, and one fixth part of a thin
Space

Space high, becaufe the Top above the *Counter-Punch*, and the Footing below, makes five fixth Parts of a thin Space, as aforefaid.

Therefore, to meafure off the Width of four thin Spaces and one fixth Part of a thin Space, lay three thin Spaces, or, which is better, the Letter e, which is three thin Spaces, as aforefaid; and . which is one thin Space and one fixth part of a thin Space, upon one another; for they make together, four thin Spaces, and one fixth part of a thin Space; and the thicknefs of thefe two Meafures fhall be the Heighth of the *Counter-Punch*, between the Footing and the Inner Angle of *A*. And thus, by this Example, you may couple with proper Meafures either the whole Forty two, which is the whole Body, or any number of its Parts, as I told you before.

This Meafure of four thin Spaces and one fixth part of a thin Space is not a Meafure, perhaps, ufed more in the whole Set of Letters to be Cut to the prefent Body, therefore you need not make a *Standing Gage* for it; yet a prefent *Gage* you muft have: Therefore ufe the *Sliding Gage* (defcribed in §. 12. ¶. 4. and *Plate* 10. at B.) and move the Socket *c c* on the Beam *a a*, till the Edge of the Sholder of the Square of the Socket at the under-fide of the Beam ftands juft the Width of four thin Spaces and one fixth part of a thin Space, from the Point of the Tooth *b*; which you may do by applying the Meafure aforefaid juft to the Square and Point of the Tooth; for then if you Screw down the Screw in the upper fide of the Sliding Socket, it will faften the Square at that diftance from the Point of the Tooth.

Tooth. And by again applying the fide of the Square to the Foot of the Face of the *Counter-Punch*, you may with the Tooth defcribe a fmall race, which will be the exact Heighth of the *Counter-Punch* for *A*. But *A* hath a Fine ftroak within it, reaching from Side to Side, which by the large Draft of *A*, you may find that the middle of this crofs ftroak is two Thin Spaces above the bottom of this *Counter-Punch*; and with your common *Sliding-Gage* meafure that diftance as before, and fet off that diftance alfo on the Face of your *Counter-Punch*. Then with the edge of a Fine *Knife-File*, File ftraight down in that race, about the depth of a Thin Space, or fomewhat more; So fhall the *Counter-Punch* for *A* be finifht. But you may if you will, take off the Edges or Sholder round about the Face of the *Counter-Punch*, almoft fo deep as you intend to ftrike it into the *Punch*: for then the Face of the *Counter-Punch* being Filed more to a Point, will eafier enter the *Punch* than the broad Flat-Face.

But note, That if it be a very Small Bodied *A* you would make, the Edge of a Thin *Knife-File* may make too wide a Groove: In this cafe you muft take a peece of a well-Temper'd broken Knife, and ftrike its Edge into the Face of the *Counter-Punch*, as aforefaid.

¶. 3. *Of Sinking the* Counter-Punches.

Having thus finifht his *Counter-Punch*, he Hardens and Tempers it, as was taught *Numb.* 3. *fol.* 57, 58. *Vol.* I. And having alfo Filed the Face of his Punch
he

he intends to cut his *A* upon, pretty Flat by guefs, he Screws the Punch upright, and hard into the Vice: And fetting the Face of his *Counter-Punch* as exactly as he can, on the middle of the Face of his Punch, he, with an Hammer fuitable to the Size of his *Counter-Punch*, ftrikes upon the end of the *Counter-Punch* till he have driven the Face of it about two Thin Spaces deep into the Face of the Punch. So fhall the *Counter-Punch* have done its Office.

But if the Letter to be *Counter-Puncht* be large, as *Great Primmer*, or upwards, I take a good high Blood red Heat of it, and Screw it quickly into the Vice; And having my *Counter-Punch* Hard, not Temper'd, becaufe the Heat of the Punch foftens it too faft: And alfo having before-hand the *Counter-Punch* Screwed into the *Hand-Vice* with its Shank along the Chaps, I place the Face of the *Counter-Punch* as before, on the middle of the Face of the Punch, and with an Hammer drive it in, as before.

Taking the Punch out of the Vice, he goes about to Flat and Smoothen the Face in earneft; for it had been to no purpofe to Flat and Smoothen it exactly before, becaufe the Sinking of the *Counter-Punch* into it, would have put it out of Flat again.

But before he Flats and Smoothens the Face of the Punch, He Files by guefs the fuperfluous Steel away about the Face of the Letter, *viz.* fo much, or near fo much, as is not to be ufed when he comes to finifh up the Letter, as in this prefent Letter *A*, which ftanding upon a Square Face on the Punch, meets in an Angle at the Top of the Letter. Therefore the Sides of that Square muft be Filed away
to

to an Angle at the Top of the Face of the Punch. But
great care muſt be taken, that he Files not more away
than he ſhould: For he conſiders that the left hand
Stroak of *A* is a Fat Stroak, and that both the
left-hand and the right-hand Stroak too, have Foot-
ings, which he is careful to leave Steel enough in
their proper places for.

The reaſon why theſe are now Fil'd thus away,
and not after the Letter is finiſht, is, Becauſe in the
Flatting the Face there is now a leſs Body of Steel
to File away, than if the whole Face of the Punch
had remain'd intire: For though the following ways
are quick ways to Flatten the Face, yet conſidering
how tenderly you go to Work, and with what
Smooth Files this Work muſt be done, the riddance
made will be far leſs when a broad Face of *Steel* is
to be Flatned, than when only ſo much, or very
little more than the Face of the *Letter* only is to be
Flatned.

To Flat and Smoothen the Face of the Punch,
he uſes the *Flat-Gage*, (deſcribed §. 12. ¶. 3. and
Plate 10. at A.) thus, He fits one convex corner of
the Shank of the Punch, into the Concave corner
of the *Flat-Gage*, and ſo applies his *Flat-Gage-
Punch* and all to the Face of the *Uſing-File*, and
lets the *Counter-Puncht* end, *viz.* the Face of the
Punch Sink down to the Face of the *Uſing-File*:
And then keeping the convex Corner of the Shank
of the Punch cloſe and ſteddy againſt the Concave
corner of the *Flat-Gage*, and preſſing with one of his
Fingers upon the then upper end of the Punch, *viz.*
the Hammer-end, he alſo at the ſame time, preſſes
the

the lower end of the Punch, *viz.* The Face againft
the *Ufing-File*, and thrufts the *Flat-Gage* and *Punch*
in it fo oft forwards, till the extuberant Steel on
the Face, be Rub'd or Fil'd away: which he knows
partly by the alteration of colour and Fine Furrows made by the *Ufing-File* on the Face of the
Punch, and partly by the falling away of the parts
of the Face that are not yet toucht by the *Ufing-
File*: So that it may be faid to be truly Flat: which
he knows, when the whole Face of the Punch
touches upon the Flat of the *Ufing-File*, or at leaft,
fo much of the Face as is required in the Letter:
For all Counter-Puncht-Letters, as aforefaid, muft
have a greater Face of Steel than what the bare
Letter requires: for the reafon aforefaid.

Another way I ufe is thus. After I have Fil'd the
Face as true as I can by guefs, with a *Rough-Cut-File*,
I put the Punch into an Hand-Vice, whofe Chaps
are exactly Flat, and ftraight on the upper Face, and
fink the Shank of the Punch fo low down in the
Chaps of the Hand-Vice, that the low fide of the
Face of the Punch may lye in the fame Plain with
the Chaps; which I try with the Liner. For the
Liner will then fhew if any of the Sides ftand higher
than the Plain of the Chaps: Then I Screw the
Punch hard up, and File off the rifing fide of the
Punch, which brings the Face to an exact Level:
For the Face of the Chaps being Hard Steel, a File
cannot touch them, but only take off the aforefaid
Rifing parts of the Face of the Punch, till the
Smooth-File has wrought it all over exactly into the
fame Plain with the Face of the Chaps of the *Hand-
Vice*.　　　　　　　　　　　　　　　　　　Some

Some *Letter-Cutters* work them Flat by Hand, which is not only difficult, but tedious, and at the beſt, but done by gueſs.

The inconvenience that this Tool is ſubject to, is, That with much uſing its Face will work out of Flat. Therefore it becomes the Workman to examine it often, and when he finds it faulty to mend it.

When they *File* it Flat by Hand, they Screw the Shank of the Punch perpendicularly upright into the Chaps of the Vice, and with a *Flat-Baſtard-Cut-File*, of about Four Inches long, or if the Punch be large, the File larger, according to diſcretion, and File upon the Face, as was ſhewn *Numb.* I. *fol.* 15, 16. Then they take it out of the Vice again, and holding up the Face Horizontally between the Sight and the Light, examine by nice obſerving whether none of its Angles or Sides are too high or too low. And then Screwing it in the Vice again, as before, with a *Smooth-Cut-File*, he at once both Files down the Higher Sides or Angles, and Smoothens the Face of the Punch. But yet is not this Face ſo perfectly Flatned, but that perhaps the middle of it riſes more or leſs, above the Sides: And then he Screws it in his *Hand-Vice*, and leans the Shank of the Punch againſt the Tach, pretty near upright, and ſo as he may beſt command it, and with a *Watch-Makers Half-Round-Sharp-Cut-File*, Files upon it with the Flat-Side of his File; But ſo that he ſcarce makes his forward and backward Stroaks longer than the breadth of the Face of his Punch, leſt in a long Stroak, the hither or farther end of his File ſhould Mount or Dip, and

there-

therefore keeps his File, with the Ball of his Finger upon it, cloſe to the Face of the Punch. Then with the Liner he examines how Flat the Face of the Punch is, and if it be not yet Flat, as perhaps it will not be in ſeveral Trials, he again reiterates the laſt proceſs with the *Small-Half-Round-File*, till it be Flat. But he often Files croſs the Furrows of the File, as well becauſe it makes more riddance, as becauſe he may better diſcern how the File bears on the Face of the Punch.

When it is Flat, he takes a Small well-worn Half-Round-File, and working (as before) with the *Sharp-Cut-File*, he Smoothens the Face of the Punch.

Having thus Flatted the Face of the Punch, and brought the Letter to ſome appearance of Form, He Screws the Punch in the Hand-Vice, but not with the Shank perpendicular to the Chaps, but ſo as the Side he intends to File upon may ſtand upwards and aſlope too, and make an Angle with the Chaps of the Hand-Vice. And holding the Hand-Vice ſteddy in his left hand, he reſts the Shank of the Punch pretty near its Face upon the Tach: and then with a ſmall *Flat-File*, called a *Pillar-File*, in his right hand, holding the Smooth Thin Side of it towards the Footing of the Stem, he Files that Stem pretty near its due Fatneſs, and ſo by ſeveral reiterated proffers, leſt he ſhould File too much of the Stem away, he brings that Stem at laſt to its true Fatneſs. Then he meaſures with the Aſcending *Face-Gage*, the Heighth of the Letter: For though the *Counter-Punch* was imagin'd

(as

(as aforeſaid) to be made to an exact Heighth for
the inſide of the Letter; yet with deeper or ſhallower Sinking it into the *Punch*, the inſide oft
proves higher, or lower: Becauſe, as aforeſaid, the
Superficies of the Face of the *Counter-Punch* is leſs
than the true meaſure. But as it runs Sholdering into the Shank of the *Counter-Punch* the Figure or
Form of the inſide becomes bigger than the inſide
of the Letter ought to be. Therefore the deeper
this Sholdering Shank is ſunk into the Face of the
Punch, the higher and broader will the Form of the
inſide of the Letter be, and the ſhallower it is
Sunk in, the Shorter and Narrower by the Rule of
Contraries.

He meaſures, as I ſaid, with the Aſſending Face-
Gage, and by it finds in what good Size the Letter
is. If it be too high, as moſt commonly it is, becauſe the Footing and Top are yet left Fat, then
with ſeveral proffers he Files away the Footing and
Top, bringing the Heighth nearer and nearer ſtill,
conſidering in his Judgment whether it be proper-
eſt to File away on the Top or Footing, till at laſt
he fits the Heighth of the Letter by the Aſſending
Face-Gage.

But though he have fitted the Heighth of the
Letter, yet if the *Counter-Punch* were made a little
too little, or Sunk a little too ſhallow, not only the
Footing will prove too Fat, but the Triangle above
the Croſs-ſtroke of *A* will be too ſmall; or if too
big, the Footing and part of the Top will be Filed
away, when it is brought to a due Heighth, and
then the Letter is Spoil'd, unleſs it be ſo deep Sunk,
that

that by working away the Face, as aforefaid, he can regain the Footing and Top through the Slope-fholdering of the *Counter-Punch*, and alfo keep the infide of the Letter deep enough.

But if the Footing be too *Fat* or the Triangle of the Top too little in the Infide, he ufes the Knife-backt Sculpter, and with one of the edges or both, that proceeds from the Belly towards the Point of the Sculpter (which edges we will for diftinⅆion fake call *Angular edges*) he by degrees and with feveral proffers Cuts away the Infide of the Footing, or opens the Triangle at the Top or both, till he hath made the Footing lean enough, and the Triangle big enough.

But if he works on the Triangle of the Top, he is careful not to Cut into the Straight of the Infide lines of the Stems, but to keep the Infides of that Triangle in a perfeⅆ ftraight line with the other part of the Infide of the Stem.

The fmall arch of a Circle on the Top of *A* is Fil'd away with a Sizable Round-File. And fo for all other Letters that have Hollows on their Outfides; he fits himfelf with a fmall File of that fhape and Size that will fit the Hollow that he is to work upon: For thus the Tails of Swafh Letters in Italick Capitals are Fil'd with half Round Files Sizable to the Hollows of them. But I inftead of Round or Half-Round Files, in this Cafe, befpeak Pillar Files of feveral Thickneffes, and caufe the *File-maker* to Round and Hatch the Edges: which renders the File ftrong and able to endure hard leaning on, without Breaking, which Round or Half-Round Files will not Bear.

I need

I need give no more Examples of Letters that are to be Counter-punched: And for Letters that need neither Counter-punching or Graving, they are made as the Out-fides of *A*, with Files proper to the fhapes of their Stroaks.

¶. 4. *Of Graving and Sculping the Infides of* Steel Letters.

The *Letter-Cutter* elects a *Steel Punch* or *Rod*, a fmall matter bigger than the Size of the Letter he is to Cut; becaufe the Topping or Footing Stroaks will be ftronger when they are a little Bevell'd from the Face. The Face of thefe Letters not being to be Counter-punched are firft Flatned and Smoothed, as was fhewed, ¶. 3. Then with the proper *Gage*, *viz.* the Long, the Afcending, or elfe the Short *Face-Gage*, according as the Letter is that he intends to Cut, He meafures off the exact Heighth of the Letter, Thus; He firft Files one of the Sides of the Face of the *Punch* (*viz.* that Side he intends to make the Foot of his Letter) exactly ftraight; which to do, he fcrews his *Punch* pretty near the bottom end, with its intended Foot-fide uppermoft, aflope into one end of the Chaps of his *Hand-Vice.* So that the Shank of the *Punch* lies over the Chaps of the *Hand-Vice,* and makes an Angle of about 45 De-grees with the Superficies of the Chaps of it: Then he lays the under fide of the Shank of his *Punch* aflope upon his *Tache,* in one of the Notches of it, that will beft fit the fize of his *Punch,* to keep it fteady; and fo Files the Foot-Line of the *Punch.*

But

9

But he Files not athwart the fides of his *Punch*;
for that might make the Foot-Line Roundifh, by a
Mounting and Dipping the Hand is prone to; as I
fhewed, *Vol.* I. *Fol.* 15, 16. But he holds his File
fo as the Length of it may hang over the Length of
the Shank of the *Punch*, and dip upon it at the Face
of the *Punch*, with a Bevel, or Angle, of about 100
Degrees with the Face of the *Punch*. This Angle
you may meafure with the *Beard-Gage*, defcribed
in *Plate* 10. *Fig.* C. at *k*. Then Filing with the File
in this Pofition, the Foot-Line will be made a true
ftraight Line. But yet he examines it too by applying the *Liner* to it; and holding the *Punch* and *Li-*
ner thus to the Light; If the *Liner* touches all the way
on the Foot-Line, he concludes it true; if not, he
mends it till it do.

Then he ufes his proper *Steel-Gage*, and places the
Sholder of it againft the Shank of the *Punch* at the
Foot-Line; and preffing the Sholder of the *Steel-*
Gage clofe againft the Foot-Line, he, with the Tooth
of the *Gage* makes a Mark or Race on the fide of
the Face, oppofite to the Foot-line: And that Mark
or Race fhall be from the Foot-Line, the Bounds of
the Heighth of that Letter.

Then on the Face he draws or marks the exa&
fhape of the Letter, with a Pen and Ink if the Letter be large, or with a fmooth blunted Point of a
Needle if it be fmall: Then with fizable and proper
fhaped and Pointed Sculptors and Gravers, digs or
Sculps out the Steel between the Stroaks or Marks
he made on the Face of the *Punch*, and leaves the
Marks ftanding on the Face.

If

If the Letter be great he is thus to Sculp out, he then, with a Graver, Cuts along the Infides of the drawn or marked Stroaks, round about all the Hollow he is Cutting in. And having Cut about all the fides of that Hollow, he Cuts other ftraight Lines within that Hollow, clofe to one another (either parallel or aflope, it matters not) till he have filled the Hollow with ftraight Lines; and then again, Cuts in the fame Hollow, athwart thofe ftraight Lines, till he fill the Hollow with Thwart Lines alfo. Which ftraight Lines, and the Cuttings athwart them, is only to break the Body of Steel that lies on the Face of the *Punch* where the Hollow muft be; that fo the Round-Back'd Sculptor may the eafier Cut through the Body of the Steel, in the Hollow, on the Face of the *Punch*; even as I told you, *Numb.* 4. *Vol.* I. §. 2. the Fore-Plain makes way for the Fine Plains.

The *Letter-Cutter* does not expeft to perform this Digging or Sculping at one fingle Operation; but, having brought the Infide of his Letter as near as he can at the firft Operation, he, with the flat fide of a Well-worn, Small, Fine-Cut, Half-round File, Files off the Bur that his Sculptors or Gravers made on the Face of the Letter, that he may the better and nicelier difcern how well he has begun. Then he again falls to work with his Sculptors and Gravers, mending, as well as he can, the faults he finds; and again Files off the Bur as before, and mends fo oft, till the Infide of his Letter pleafes him pretty well. But before every Mending he Files off the Bur, which elfe, as aforefaid, would obfcure and hide the true fhape of his Stroaks.   Having

Having well fhaped the Infide Stroaks of his Letter, he deepens the Hollows that he made, as well
as he can, with his Sculptors and Gravers: And the
deeper he makes thefe Hollows, the better the Letter will prove. For if the Letters be not deep enough,
in proportion to their Width, they will, when the
Letter comes to be Printed on, Print Black, and
fo that Letter is fpoiled.

How deep thefe Hollows are to be, cannot be
well afferted, becaufe their Widths are fo different,
both in the fame Letter, and in feveral Letters:
Therefore he deepens them according to his Judgment and Reafon. For Example, O muft be deeper
than A need be, becaufe the Hollow of O is wider
than the Hollow of A; A having a Crofs Stroak in
it; and the wider the Hollow is, the more apt will
the wet Paper be to prefs deeper towards the bottom in Printing. Yet this in General for the Depth
of Hollows; You may make them, if you can, fo
deep as the *Counter-Punch* is directed to be ftruck
into the Face of the Punch. See ¶. 3. of this §.

Having with his Gravers and Sculptors deepned
them fo much as he thinks convenient, he, with a
Steel Punch, pretty near fit to the fhape and fize of
the Hollow, and Flatted on its Face, Flattens down
the Irregularities that the Gravers or Sculptors made,
by ftriking with a proper Hammer, upon the Hammer-end of the *Punch*, with pretty light blows. But
he takes great care, that this *Flat-Punch* be not at
all too big for the Hollow it is to be ftruck into, left
it force the fides of the Stroaks of the Letter out of
their fhape: And therefore alfo it is, that he ftrikes
                                                    but

but eafily, though often, upon the end of the *Flat-Punch*.

Having finifhed the Infide, he works the Outfides with proper Files; as I fhewed before, in Letter A; and fmoothens and Pollifhes the Outfide Stroaks and Face with proper worn-out fmall *Watch-makers* Files.

The Infide and Outfide of the Face thus finifhed, he confiders what Sholdering the Shank of the *Punch* makes now with the Face, round about the Letter. For, as the Shank of the Letter ftands farther off the Face of any of the Stroaks, the Sholdering will be the greater when the Letter is firft made; becaufe the Outfides of the Letter, being only fhaped at firft with Fine Small Files, which take but little Steel off, they are Cut Obtufely from the Shank to the Face, and the Steel of the Shank may with Rougher Files afterwards, be Cut down more Tapering to the Shank.   For the Sholder of the Shank, as was faid before in this ¶, muft not make an Angle with the Face, of above 100 Degrees; becaufe elfe they would be, firft, more difficult to Sink into Copper; And Secondly, The broad Sholders would more or lefs (when the Letter is Caft in fuch Matrices) and comes to the Prefs, be fubject, and very likely to be-fmear the Stroaks of the Letter; efpecially, with an Hard Pull, and too wet Paper; which fqueezes the Face of the Letter deep into the Paper, and fo fome part of the Broad Sholdering of the Letter, receiving the Ink, and preffing deep into the Paper, flurs the Printed Paper, and fo makes the whole Work fhew very nafty and un-beautiful.

For

9*

For thefe Reafons it is, that the Shank of the *Punch*, about the Face, muft be Filed away (at leaft, fo much as is to be Sunk into Copper) pretty clofe to the Face of the Letter; yet not fo as to make a Right Angle with the Face of the Letter, but an Obtufe Angle of about 100 Degrees: For, fhould the Shank be Filed away to a Right Angle, *viz.* a Square with the Face, if any Footing or Topping be on the Letter, thefe fine Stroaks will be more fub-ject to break when the *Punch* is Sunk into Copper, than when the Angle of the Face and Shank is aug-mented; becaufe then thofe fine Stroaks ftand upon a ftronger Foundation. Therefore he ufes the *Beard-Gage*, and with that examines round about the Let-ter, and makes the Face and Shank comply with that.

Yet Swafh-Letters, efpecially ℒ, whofe Swafhes come below the Foot-Line, and whofe Length reaches under the Foot-Line of the next Letter, or Letters in Compofing, ought to have the Upper Sholder of that Swafh Sculped down ftraight, *viz.* to a Right Angle, or Square with the Face; at leaft, fo much of it as is to be Sunk into Copper: Becaufe the Upper Sholder of the Swafh would elfe be fo broad, that it would ride upon the Face of the next Letter. Therefore the Swafh-Letters being all Long Letters, the lower end of the Swafhes reach as low as the Bottom-Line; which cannot be Filed Square enough down from the Head-Line, unlefs the Steel the Swafh ftands on, fhould be Filed from end to end, the length of the whole Shank of the *Punch*, which would be very tedious; and befides, would

make

make that part of the Shank the Swaſh ſtands on ſo weak, that it would ſcarce endure Striking into the Copper. Therefore, as I ſaid before, the Upper Sholder of the Swaſh ought to be Sculped down: Yet I never heard of any *Letter-Cutters* that had the knack of doing it; but that they only Filed it as ſtraight down as they could, and left the *Letter-Kerner*, after the Letter was Caſt, to Kern away the Sholdering. Yet I uſe a very quick way of doing it; which is only by Reſting the Back of a Graver at firſt, to make way; and afterwards a Sculptor, upon the Shank of the *Punch*, at the end of the Swaſh, one while; and another while on the Shank, at the Head, that the Swaſh may be Sculped down from end to end: and Sculping ſo, Sculp away great Flakes of the Steel at once, till I have Cut it down deep enough, and to a Right Angle.

Then he Hardens and Tempers the Punch; as was ſhewed, *Numb.* 3. *Vol.* I. *Fol.* 57, 58.

But though the *Punch* be Hardned and Temper'd, yet it is not quite finiſhed: for, in the Hardning, the *Punch* has contraƈted a Scurf upon it; which Scurf muſt be taken off the Face, and ſo much of the ſides of the Shank as is to be Sunk into Copper. Some *Letter-Cutters* take this Scurf off with ſmall ſmooth Files, and afterwards with fine Powder of *Emerick*. The *Emerick* they uſe thus. They provide a Stick of Wood about two Handful long, and about a *Great-Primer*, or *Double-Pica* thick: Then in an Oyſter-ſhell, or any ſleight Concave thing, they powr a little Sallad-Oyl, and put Powder of *Emerick* to it, till it become of the Conſiſtence of Batter
made

made for Pan-cakes. And ſtirring this Oyl and *Emerick* together, ſpread or ſmear the aforeſaid Stick with the Oyl and *Emerick*, and ſo rub hard upon the Face of the *Punch*, and alſo upon part of the Shank, till they have taken the Scurf clean off.

Mr. *Walberger* of *Oxford* uſes another way. He makes ſuch an Inſtrument as is deſcribed in *Plate* 10. at H, which we will, for diſtinction ſake, call the *Joynt-Flat-Gage*. This Inſtrument conſiſts of two Cheeks about nine Inches long, as at b, and are faſtned together at one end, as the Legs of a *Carpenter*'s Joynt-Rule are in the Centre, as at *c*, but with a very ſtrong Joynt; upon which Centre, or Joynt, the Legs move wider, or cloſer together, as occaſion requires. Each Leg is about an Inch and a quarter broad, and an Inch and three quarters deep; *viz.* ſo deep as the Shank of the *Punch* is long. At the farther end of the Shank b (as at *d*) is let in an Iron Pin, with an Head at the farther end, and a ſquare Shank, to reach almoſt through a ſquare Hole in the Shank b, that it twiſts not about; and at the end of that Square, a round Pin, with a Male-Screw made on it, long enough to reach through the Shank a, and about two Inches longer, as at *e*; upon which Male-Screw is fitted a Nut with two Ears, which hath a Female-Screw in it, that draws and holds the Legs together, as occaſion requires a bigger or leſs *Punch* to be held in a proper Hole. Through each of the adjoyning Inſides of the Legs are made, from the Upper to the Lower Side, ſix, ſeven, or eight Semi-Circular Holes (or more or leſs, according to diſcretion) exactly Perpendicular
to

to the upper and under Sides of each Leg, marked
a a a a, b b b b.   Each of thefe Semicircular Holes is,
when joyned to its Match, on the other Leg to
make a Circular Hole; and therefore muſt be made
on each Leg, at an equal diſtance from the Centre.
Thefe Holes are not all of an equal Size, but different Sizes: Thofe towards the Centre fmalleſt, *viz.*
fo fmall, that the *Punch* for the fmalleſt Bodied Letters may be pinched faſt in them; and the biggeſt
Holes big enough to contain, pinch and hold faſt
the *Punches* for the great Bodied Letters.   The upper and under fides of this *Joynt-Flat-Gage* is Faced
with an Iron Plate, about the thickneſs of an Half
Crown, whofe outer Superficies are both made exactly Flat and Smooth.

When he ufes it, he chufes an Hole to fit the Size
of the *Punch*; and putting the Shank of the *Punch*
into that Hole, Sinks it down fo low, till the Face
of the *Punch*, ſtands juſt Level, or rather, above
the Face of the *Joynt-Flat-Gage:* Then with a piece
of an Hone, wet in Water, rubs upon the Face of
the *Punch*, till he have wrought off the Scurf.   At
laſt, with a Stick and Dry *Putty*, Poliſhes it.

I like my own way better than either of the former: For, to take off the Scurf with Small Files
fpoils the Files; the Face of the *Punch* being Hard,
and the Scurf yet Harder: And befides, endangers
the wronging the Face of the *Punch*.

The *Joynt-Flat-Gage* is very troublefome to ufe,
becaufe it is difficult to fit the Face of the *Punch*, to
lie in the Plain of the Face of the *Gage*; efpecially,
if, in making the Letter, the Shank be Filed Tapering,

ing, as it moft times is. For then the Hammer-end of the *Punch* being bigger than the Face-end, it will indeed Pinch at the Hammer-end, whilft the Face-end ftands unfteady to Work on. But when the *Punch* is fitted in, it is no way more advantagious for Ufe, than the Chaps of the *Hand-Vice* I mentioned in ¶. 3. of this §.

Wherefore, I fit the *Punch* into the Chaps of the *Hand-Vice*, as I fhewed in the aforefaid ¶. and with a fine fmooth Whet-ftone and Water, take the Scurf lightly off the Face of the *Punch*; and afterwards, with a fine fmooth Hone and Water, work down to the bare bright Steel. At laft, drying the *Punch* and Chaps of the *Hand-Vice* with a dry Rag, I pollifh the Face of the *Punch* with Powder of Dry *Brick* and a Stick, as with *Putty.*

¶. 5. *Some Rules he confiders in ufing the* Gravers, Sculptors, Small Files, *&c.*

1. When he is Graving on the Infide of the Stroak, either to make it Finer or Smoother, he takes an efpecial care that he place his Graver or Sculptor fo, as that neither of its Edges may wrong another Stroak of the Letter, if they chance (as they often do) to flip over, or off an extuberant part of the Stroak he is Graving upon. And therefore, I fay, he well confiders how he is to manage the edges of his Graver. For there is no great danger of the point of his Graver after the infide Stroaks are form'd, and the Hollows of the Letter fomewhat deepned; but in the edges there is: For the point

in

in working lies always below the Face of the Letter, and therefore can, at moſt, but ſlip below the Face, againſt the ſide of the next Stroak; but the edges lying above the Face of the Letter, may, in a ſlip, touch upon the Side and Face of the next Stroak, and wrong that more or leſs, according as the force of the Slip was greater or ſmaller. And if that Stroak it jobs againſt were before wholly finiſhed, by that job the whole Letter is in danger to be ſpoiled; at the beſt, it cannot, without Filing the Letter lower, be wrought out; which ſometimes is a great part of doing the Letter anew: For he takes ſpecial care that neither any dawk, or the leaſt extuberant bunching out be upon the inſide of the Face of the Stroak, but that the inſide of the Stroak (whether it be Fat or Lean) have its proper Shape and Proportion, and be purely ſmooth and clean all the way.

If on the inſide of the Stroak the Graver or Sculptor have not run ſtraight and ſmooth on the Stroak, but that an Extuberance lies on the Side, that Extuberance cannot eaſily be taken off, by beginning to Cut with the Edge of the Graver or Sculptor juſt where the Extuberance begins: Therefore he fixes the Point of his Graver or Sculptor in the Bottom of the Hollow, juſt under the Stroak where the Extuberance is, and leans the Edge of his Graver or Sculptor upwards; ſo as in forcing the Point of the Graver or Sculptor forwards, at the Bottom of the Hollow, the Edge of the Graver or Sculptor may ſlide tenderly along, and take along with it a very ſmall, nay, inviſible Chip of the moſt Prominent Part of the Extuberance; and ſo, by this Proceſs reitterated

ted often, he, by fmall Degrees, Cuts away the Extuberant part of the Stroak.

2. He is careful to keep his Gravers and Sculptors
always Sharp, by often Sharpning them on the Oyl-
Stone, which for that purpofe he keeps ready at hand,
ftanding on the Bench: For if a Graver or Sculptor
be not fharp, it will neither make riddance, or Cut
fmooth; but inftead of Cutting off a fmall Extuberancy, it will rather ftick at it, and dig into the Side of
the Stroak.

3. He Files very tenderly with the Small Files,
efpecially with the Knife-Files, as well becaufe they
are Thin and Hard, not Temper'd, and therefore
would fnap to pieces with fmall violence; as alfo,
left with an heavy hand he fhould take away too
much at once of that Stroak he is working upon.

§. 14. ¶. 1. *Some Rules to be obferved by the* Letter-
Cutter, *in the Cutting* Roman, Italick, *and the*
Black Englifh Letter.

1. The Stem and other Fat Stroaks of Capital *Romans* is five Parts of forty and two (the whole
Body:) Or, (which is all one) one fixth part of the
Heighth of an Afcending Letter (as all Capitals are
Afcendents) as has been faid before. *Albertus Durer*
took his Meafure from the Heighth of Capitals, and
affigned but one tenth part for the Stem.

2. The Stem, and other Fat Stroaks of Capitals
Italick, is four parts of forty and two, (the Body.)

3. The Stem, and other Fat Stroaks of Lower-
Cafe *Roman*, is three and an half parts of forty and
two, (the Body.)   4. The

4. The Stem, and other Fat Stroaks of Lower-Cafe *Italick*, is three parts of forty and two, (the Body.)

5. Of *Englifh*, the Short Letters ftand between nine parts of the Bottom-Line, and nine parts from the Top-Line; *viz.* upon three and thirty parts of forty and two, (the Body.)

6. The Stem of *Englifh* Capitals is fix parts of forty and two, (the Body.)

7. The Stem of *Englifh* Lower-Cafe Letters is four parts of forty and two, (the Body.)

¶. 2. *Of Terms relating to the Face of Letters,*
and their Explanation.

The Parts of a *Punch* are already defcribed in §. 13. ¶. 1. of this Volumne; and fo is the Body: But the feveral Terms that relate to the Face of Letters are not yet defined. Now therefore you muft note, that the Body of a Letter hath four principal Lines paffing through it (or at leaft imagined to pafs through it) at Right Angles to the Body; *viz.* The Top-Line, The Head-Line, The Foot-Line, and The Bottom-Line.

Between two of thefe Lines is contained the Heighth of all Letters.

Thefe are called *Lines*, becaufe the Tops, the Heads, the Feet and the Bottoms of all Letters (when Complicated by the *Compofitor*) ftand ranging in thefe imagin'd Lines, according as the Heighth and Depth of each refpective Letter properly requires.

The

The Long Letters are (as I told you in §. 13. ¶. 1. of this Volumne) contained between the Top and Bottom-Lines, The Afcending Letters are contained between the Top and Foot-Lines, The Defcending Letters are contained between the Head and Bottom-Lines, and The Short Letters are contained between the Head and Foot-Lines.

Through what Parts of the Body all thefe Lines pafs, you may fee by the Drafts of Letters, and the following Defcriptions.

What the Long Letters, Afcending Letters, and Short Letters are, I fhewed in the afore-cited ¶. Therefore I fhall now proceed to particular Terms relating to the Face. As,

1. The Topping, is the ftraight fine Stroak or Stroaks that lie in the Top-Line of Afcending Letters: In *Roman* Letters they pafs at Right Angles through the Stems; but in *Italicks*, at Oblique Angles to the Stems; as you may fee in the Drafts of Letters, B, *B*, H, *H*, I, *I*, &c.

2. The Footing, is the ftraight fine Stroak or Stroaks that lie in the Foot-Line of Letters, either Afcending or Defcending. In *Romans* they pafs at Right Angles through the Stem, but in *Italicks*, at Oblique Angles; as you may fee in B, *B*, H, *H*, I, *I*, &c.

3. The Bottom-Footing, is the ftraight fine Stroaks that lie in the Bottom-Line of Defcending Letters. In *Romans* they pafs at Right Angles through the Stem; but in *Italicks* at Oblique Angles; as you may fee p, *p*, q, *q*.

4. The

4. The Stem is the ſtraight Fat Stroak of the Letter: as in B, *B,* the ſtraight Stroak on the Left Hand is the Stem; and I, *I,* is all Stem, except the Footing and Topping.

5. Fat-Stroaks.   The Stem or broad Stroak in a Letter is called Fat; as the Right Hand Stroak in A, and part of the great Arch in B, are Fat Stroaks.

6. Lean Stroaks, are the narrow fine Stroaks in a Letter; as the Left Hand Stroak of A, and the Right Hand Stroak of V are Lean.

7. Beak of Letters, is the fine Stroak or Touch that ſtands on the Left Hand of the Stem, either in the Top-Line, as b d h, *&c.* or in the Head-Line, as i, m, n, *&c.*   Yet f, g, ſ, *f, g, ſ,* have Beaks on the Right Hand of the Stem.

8. Tails of Letters, is a Stroak proceeding from the Right Hand Side of the Stem, in the Foot-Line; as a d t u: and moſt *Italick* Lower-Caſe Letters have Tails: As alſo have moſt Swaſh Letters.   But ſeveral of their Tails reach down to the Bottom-Line.

9. Swaſh Letters are *Italick* Capitals; as you ſee in *Plate* 15.

Thus much of *Letter-Cutting.*   The next *Exerciſes* ſhall (God willing) be upon *Making Matrices, Making Molds, Caſting and Dreſſing of Letters,* &c.

F   I   N   I   S.

ADVERTISEMENT.

Plate 11

A B C D E

A Scale of 42 Parts Viz. the Body.

F G H I J

K L M N

O P Q R

S T V U

W X Y Z

10

Plate 12.

ÆEabcdef

A Scale of 42 Parts Viz the Body.

ghijklmn

opqrſstv

uwxyz&

&tſtſhffiæœ

1234567800

Plate 13.

ABCDE

A Scale of 42 Parts Viz. the Body.

FGHIK

LMNO

PQ RST

VWXY

Z Æ

10*

Plate 14.

a b c d e f g h i

j k l m n n o o p q

r ſ s t u v w x

y z & ct ſt ſh

æ œ ſſ ſſi ſſ ffi ffl

Plate 15.

A B C D E

A Scale of 42 Parts viz. ⅙ Body

F G H J K

M N P

Q R T

U X Y Z

Æ

Plate 16.

A Scale of 42 Parts Viz the Body

Plate 27.

a b c d e f g h i j

A Scale, of 42 Parts via the Body.

k l m n o p q r

A s t b u w x y

z t a t t h h t t

h æ œ P S * †

,, ., .?" (!)

MECHANICK EXERCISES:

Or, the Doctrine of

Handy-works.

Applied to the A R T of

Mold-Making, Sinking the *Matrices,*
Casting and Dressing of

Printing-Letters.

§. 15. ¶. 1. *Of making the* Mold.

THE *Steel Punches* being thus finish'd, as afore was shewed, they are to be sunk or struck into pieces of *Copper,* about an Inch and an half long, and one quarter of an Inch deep; but the thickness not assignable, because of the different thicknesses in Letters, as was shewed in §. 2. and shall further be shewed, when I come to the sinking and justifying

ftifying of *Matrices.* But before thefe *Punches* are
funk into *Copper*, the *Letter-Founder* muft provide a
Mold to juftifie the *Matrices* by: And therefore it is
proper that I defcribe this *Mold* to you before I proceed any farther.

I have given you in Plate 18. at A, the Draft of
one fide or half of the *Mold*; and in Plate 19. at B,
its Match, or other half, which I fhall in general thus
defcribe.

Every *Mold* is made of two parts, an under, and
an upper Part; the under part is delineated at A, in
Plate 18, the upper part is marked B, in Plate 19,
and is in all refpects made like the under part, excepting the *Stool* behind, and the *Bow*, or *Spring* alfo behind; and excepting a fmall roundifh *Wyer* between
the *Body* and *Carriage*, near the *Break*, where the
under part hath a fmall rounding *Groove* made in the
Body. This *Wyer*, or rather *Half-Wyer* in the upper
part makes the *Nick* in the *Shank* of the Letter, when
part of it is received into the Grove in the under part.

Thefe two parts are fo exactly fitted and gaged into one another, (*viz.* the *Male Gage*, marked C in
Plate 19, into the *Female-Gage* marked g, in Plate
18.) that when the upper part of the *Mold* is properly placed on, and in the under part of the *Mold*
both together, makes the entire *Mold*, and may be
flid backwards for Ufe fo far, till the Edge of either
of the Bodies on the middle of either *Carriage* comes
juft to the Edge of the *Female-Gages*, cut in each *Carriage*: And they may be flid forwards fo far, till the
Bodies on either *Carriage* touch each other. And the
fliding of thefe two parts of the *Mold* backwards,
makes

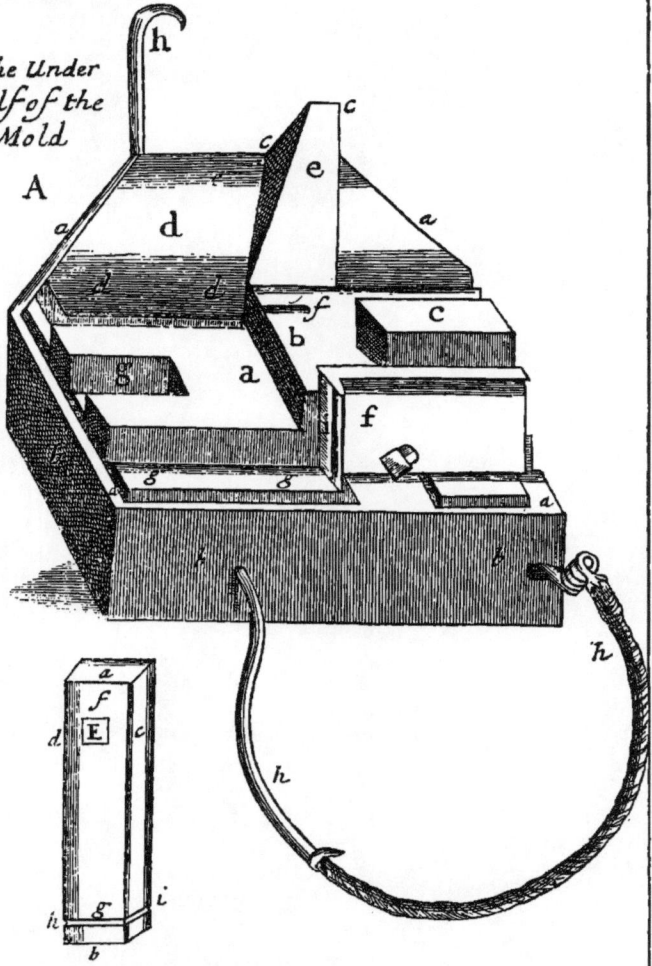

Plate 18

The Under half of the Mold

makes the *Shank* of the Letter thicker, becaufe the
Bodies in each part ftand wider afunder; and the fliding them forwards makes the *Shank* of the Letter
thinner, becaufe the Bodies on each part of the *Mold*
ftand clofer together.

This is a general Defcription of the *Mold*; I come
now to a more particular Defcription of its parts.

a The *Carriage.*
b The *Body.*
c The *Male-Gage.*
d e The *Mouth-Piece.*
f i The *Regifter.*
g The *Female-Gage.*
h The *Hag.*
a a a a The *Bottom Plate.*
b b b The *Wood* the *Bottom Plate* lies on.
c c e The *Mouth.*
d d The *Throat.*
e d d The *Pallat.*
f The *Nick.*
g g The *Stool.*
h h g The *Spring* or *Bow.*

I have here given you only the Names of the parts
of the *Mold,* becaufe at prefent I purpofe no other
Ufe of it, than what relates to the finking the *Punches* into the *Matrices*: And when I come to the cafting of Letters, You will find the Ufe and Neceffity
of all thefe Parts.

¶ 2. *Of*

¶ 2. *Of the* Bottom-Plate.

The *Bottom Plate* is made of *Iron*, about two Inches and three quarters long, and about the same breadth; its thickness about a *Brevier*: It is planisht exactly flat and streight: It hath two of its *Fore-Angles*, as *a a* cut off either straight or rounding, according to the pleasure of the Work-man.

About the place where the middle of the *Carriage* lies, is made a Hole about a *Great Primmer* square, into which is rivetted on the upper side a Pin with a Sholder to it, which reaches about half an Inch through the under side of the *Bottom Plate*. This *Pin* on the under side the *Bottom Plate* is round, and hath a *Male-Screw* on its end. This *Pin* is let through a Hole made in the Wood of the *Mold* to fit it; so that when a square *Nut*, with a *Female-Screw* in it, is turned on the *Male-Screw*, it may draw and fasten the *Half Mold* firm to the Wood.

The Hind side of the *Carriage* lies on this *Bottom-Plate*, parallel to the Hind side of it, and about a *Two-Lin'd-English* within the Hind Edge of it; and so much of this *Bottom-Plate* as is between the *Register* and the left hand end of the *Carriage* (as it is posited in the Figure) is called the *Stool*, as *g g* in the under half of the *Mold*, because on it the lower end of the *Matrice* rests; but on the upper half of the *Mold* is made a square Notch behind in the *Bottom-Plate*, rather within than without the Edge of the *Carriage*, to reach from the *Register*, and half an Inch towards the left hand (as it is posited in the Figure) that

that the upper part of the fore-fide of the *Matrice* may ftand clofe to the *Carriage* and *Body.*

¶ 3. *Of the* Carriage.

On the *Bottom-Plate* is fitted a *Carriage,* (as *a*) This *Carriage* is almoft the length of the *Bottom-Plate,* and about a *Double Pica* thick, and its Breadth the length of the Shank of the Letter to be caft.

This *Carriage* is made of *Iron,* and hath its upper fide, and its two narrow fides filed and rubed upon the ufing *File,* exactly ftraight, fquare and fmooth, and the two oppofite narrow fides exactly parallel to each other.

On one end of the *Carriage,* as at g, is made a long *Notch* or *Slit,* which I call the *Female-Gage:* It is about a *Double Pica* wide, and is made for the *Male-Gage* of the other part of the *Mold* to fit into, and to flide forwards or backwards as the thicknefs of the Letter to be caft may require.

¶ 4. *Of the* Body.

Upon the *Carriage* is fitted the *Body,* as at b. This *Body* is alfo made of *Iron,* and is half the length of the *Carriage,* and the exact breadth of the *Carriage*; but its thicknefs is alterable, and particularly made for every intended *Body.*

About the middle of this *Body* is made a fquare Hole, about a *Great Primmer,* or *Double Pica* fquare; and directly under it is made through the *Carriage* fuch another Hole exactly of the fame fize.

¶ 5. *Of*

11*

¶ 5. *Of the* Male-Gage.

Through thefe two Holes, *viz.* That in the *Body*,
and that in the *Carriage,* is fitted a fquare *Iron Shank*
with a *Male-Screw* on one End, and on the other
End an Head turning fquare from the fquare *Shanck*
to the farther end of the *Body*, as is defcribed at c;
but is more particularly defcribed apart at B in
the fame Plate, where B may be called the *Male-
Gage*: For I know no diftinct Name that *Founders*
have for it, and do therefore coyn this:
  a The *fquare Shanck.*
  b The *Male-Screw.*
This *fquare Shanck* is juft fo long within half a *Scaboard* thick as to reach through the *Body, Carriage,* and
another fquare Hole made through the *Bottom-Plate,*
that fo when a *fquare Nut* with a *Female-Screw* in it
is turned on that *Pin*, the *Nut* fhall draw and faften
the *Body* and *Carriage* down to the *Bottom-Plate.*
  The Office of the *Male-Gage* is to fit into, and flide
along the *Female-Gage.*

¶ 6. *Of the* Mouth-Piece.

Clofe to the *Carriage* and *Body* is fitted a *Mouth-
Piece* marked d e.   *Letter-Founders* call this altogether a *Mouth-Piece*: But that I may be the better un-
derftood in this prefent purpofe, I muft more nicely
diftinguifh its parts, and take the Freedom to elect
Terms for them, as firft,
  c c e The *Mouth.*

d The

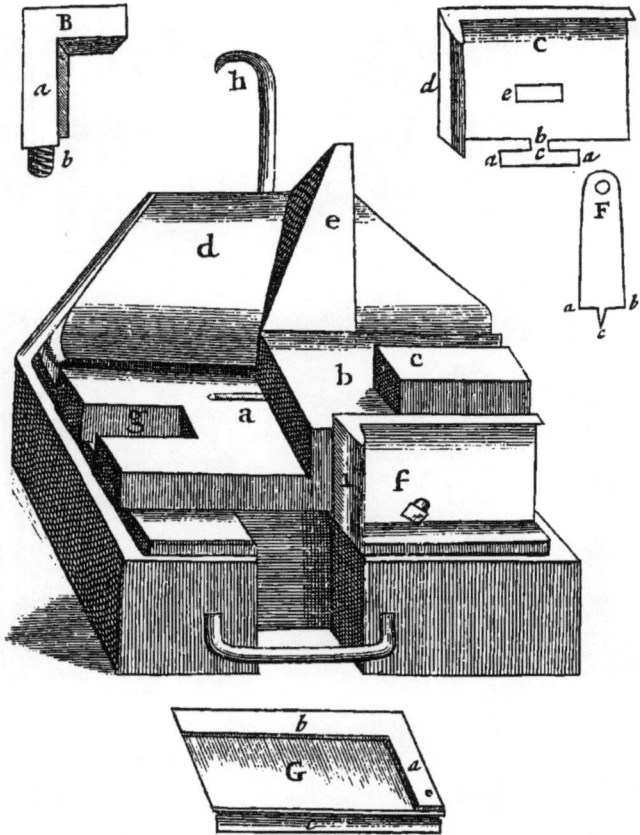

Plate 19.

The Upper half of the Mold

d The *Palate*.

c c e d The *Jaws*.

d d The *Throat*.

Altogether (as aforefaid) the *Mouth-Piece*.

The *Mouth-Piece* hath its *Side* returning from the *Throat* filed and rubb'd on the *Ufing File* exactly ftraight and fquare to its *Bottom-fide*, becaufe it is to joyn clofe to the Side of the *Carriage* and *Body*; but its upper Side, *viz.* the *Palate* is not parallel to the *Bottom*, but from the Side *d d*, *viz.* the *Throat* falls away to the *Mouth* e, making an *Angle* greater or fmaller, as the *Body* that the *Mold* is made for is bigger or lefs: For fmall *Bodies* require but a fmall *Mouth*, becaufe fmall *Ladles* will hold Metal enough for fmall Letters; and the fmaller the *Ladle*, the finer the *Geat* of the *Ladle* is; and fine *Geats* will eafier hit the *Mouth* (in a Train of Work) than the courfe *Geats* of Great *Ladles*: Therefore it is that the *Mouth* muft be made to fuch a convenient Width, that the *Ladle* to be ufed and its *Geat*, may readily, and without flabbering, receive the Metal thrown into the *Mold*.

But again, if the *Mouth-Piece* be made too wide, *viz.* the *Jaws* too deep at the *Mouth*, though the *Geat* of the *Ladle* does the readier find it, yet the *Body* of the *Break* of the *Letter* will be fo great, that firft it heats the *Mold* a great deal fafter and hotter; and fecondly, it empties the *Pan* a great deal fooner of its Metal, and fubjects the Workman fometime to ftand ftill while other Metal is melted and hot: Therefore Judgment is to be ufed in the width of the *Mouth*; and though there be no Rule for the width of it; yet this in general for fuch *Molds* as I

make

make, I obferve that the *Orifice* of the *Throat* may be about one quarter of the Body for fmall Bodies; but for great Bodies lefs, according to Difcretion, and the *Palate* about an Inch and a quarter long from the *Body* and *Carriage*. The reafon that the *Orifice* of the *Throat* is fo fmall, is, becaufe the Subftance at the end of the *Shanck* of the Letter ought alfo to be fmall, that the *Break* may eafier break from the *Shanck* of the Letter, and the lefs fubject the *Shanck* to bowing; for the bowing of a Letter fpoils it; and the reafon why the *Palate* is fo long, is, that the *Break* being long, may be the eafier finger'd and manag'd in the breaking.

If it be objected, that fince the fmalnefs of the *Break* at the end of the *Shanck* of the Letter is fo approvable and neceffary for the reafon aforefaid, then why may not the *Break* be made much more fmaller yet? The Anfwer will be, No; becaufe if it be much fmaller than one quarter of the Body, Metal enough will not pafs through the *Throat*, to fill both the *Face* and *Shanck* of the Letter, efpecially if the Letter to be caft prove thin.

Near the *Throat* and *Jaw* is made ftraight down through the *Palate* a fquare Hole (as at *k*.) This fquare Hole hath all its Sides on the *Upper-Plain* of the *Palate* opened to a *Bevel* of about 45 Degrees, and about the depth of a thick *Scaboard*. Into this fquare *Hole* is fitted a fquare *Pin* to reach through it; and within half a *Scaboard* through a fquare *Hole*, made juft under it in the *Bottom-Plate* which the *Mouth-Piece* lies upon. On the upper end of this fquare *Pin* is made a fquare *Sholder*, whofe under fides are filed *Bevil* away, fo as

the

to comply and fall juft into the *Bevil* made on the *Palate* aforefaid, and on the under end of the *Pin* is made a *Male-fcrew* long enough to contain a fquare *Nut*, with a *Female-fcrew* in it about a *Pica* or *Englifh* thick, which *Nut* being twifted about the *Pin* of the *Male-fcrew*, draws and faftens the *Mouth-piece* clofe down to the *Bottom-Plate*, and alfo clofe to the *Carriage* and *Body* of the *Mold*.

Note, that the fquare *Hole* made in the *Bottom-Plate* to receive the fquare *Shanck* of the *Pin*, muft be made a little wider than juft to fit the fquare *Shanck* of the *Pin*, becaufe the *Mouth-piece* muft be fo placed, that the end of the *Jaw* next the *Throat* muft lie juft even with the *Body* it is to be joyned to; and alfo that the *Throat* of the *Mouth-piece* may be thruft perfectly clofe to the Sides of the *Carriage* and *Body*: And when Occafion requires the *Shanck* of the Letter to be lengthned, it may be fet farther off the *Carriage*, that an *Afidue*, or fometimes a thin *Plate* of *Brafs* may be fitted in between the *Carriage* and the *Throat* of the *Mouth-piece*, as fhall farther be fhewed when I come to juftifie the *Mold*.

¶ 7. *Of the* Regifter.

Behind the *Mold* is placed the *Regifter*, as at *f i h*, which I have alfo placed apart in the aforefaid Plate, as at C, that it may the more perfpicuoufly be difcerned, and a more particular account of its parts be given, which are as follows:

C *a a b c d e* The *Regifter*.
a a The *Sholders*.

b c The

b c The *Neck.*

d The *Cheek* returning square from the Plate of the *Register*, and is about an *English* thick.

e The *Screw Hole.*

It is made of an *Iron* Plate about a *Brevier* thick; its upper Side is straight, but its under Side is not: For at *a a* projects downwards a small piece of the same Plate, which we may call the *Sholders*, of the Form you see in the Figure. These *Sholders* have two small *Notches* (as at *b c*) filed in them below the *Range* on the under side of the *Register*, which we will call the *Neck*, and is just so wide as the *Bottom-Plate* is thick. This *Neck* is set into a square *Notch*, filed so far into the *Bottom-Plate*, that the flat inside of the *Register* may stand close against the hind side of the *Carriage* and *Body*; and this *Notch* is filed so wide on the left Hand, that when the side *b* of the *Neck* stands close against the left-hand Side of this *Notch* (as it is posited in the Figure) the *Cheek* of the *Register* stands just even with the Edge of the *Body*. And this *Notch* is also filed so wide on the right-Hand Side, that when the *Neck* at *c* stands close against the right-hand Side of the *Notch*, the *Cheek* of the *Register* may remove an m, or an m and an n from the edge of the *Body* towards the right hand: And the *Sholders a a* are made so long, that when either Side of the *Neck* is thrust close against its corresponding side in the *Notch* of the *Bottom-Plate*, the upper Edge of the opposite *Sholder* shall hook or bear against the under side of the *Bottom-Plate*, and keep the whole *Register* steady, and directly upright to the Surface of the *Bottom-Plate.*

In

In the Plate of the *Regiſter*, is made a long ſquare Hole, as at *e*, juſt wide enough to receive the *Pin* of a *Male-ſcrew*, with a *Sholder* to it, which is to fit into a *Female-ſcrew*, made in the Edge of the *Carriage*, that when the *Male-ſcrew* is turned about in the *Female-ſcrew* in the *Carriage*, it ſhall draw the *Sholder* of the ſaid *Male-ſcrew* hard againſt the upper and under Sides of the ſquare Hole in the Plate of the *Regiſter*, cloſe to the ſide of the *Carriage* and *Body*.

The reaſon why the Hole in the Plate of the *Regiſter* is made ſo long, is that the *Cheek* of the *Regiſter* may be ſlid forwards or backwards as occaſion requires; as ſhall be ſhewn when I come to juſtifying the *Mold*.

¶ 8. *Of the* Nick.

In the upper half of the *Mold*, at about a *Pica* diſtance from the *Throat*, is fitted into the under ſide of the *Body* the *Nick*: It is made of a piece of *Wyer* filed flat a little more than half away. This *Nick* is bigger or leſs, as the Body the *Mold* is made for is bigger or leſs; but its length is about two *m*'s. It is with round *Sculptors* let exactly into the under ſide of the *Body*.

In the under half of the *Mold*, is made at the ſame diſtance from the *Throat*, on the upper ſide of the *Body*, a round *Groove*, juſt fit to receive the *Nick* in the upper half.

¶ 9 *Of*

¶ 9. *Of the* Bow *or* Spring.

This is a long piece of hard *Iron Wyer*, whose Diameter is about a *Brevier* thick, and hath one end faftned into the Wood of the under half of the *Mold*,
as at *h*; but it is fo faftned, that it may turn about in
the Hole of the Wood it is put into: For the end of
it being batter'd flat, a fmall Hole is drilled through
it, into which fmall Hole the end of fine *Lute-ftring
Wyer*, or fomewhat bigger is put, and faftned by twift-
ing about half an Inch of the end of the *Lute-ftring*
to the reft of the *Lute-ftring*: For then a confiderable
Bundle of that *Wyer*, of about the Size of a Doublet
Button, being wound behind the Hole, about the
end of the *Spring*, will become a *Sholder* to it, and
keep the end of the *Spring* from flipping through the
Hole in the Wood: But this *Button* or *Sholder* muft
alfo be kept on by thrufting another piece of *Wyer*
ftiff into the Hole made on the end of the *Spring*,
and crooking that *Wyer* into the Form of an S, that
it flip not out of the Hole.

The manner how the *Spring* is bowed, you may
fee in the Figure: But juft without the Wood is
twifted upon another *Wyer* about an *English* thick
five or fix turns of the *Wyer* of the *Spring*, to make
the whole *Spring* bear the ftronger at its point: For
the Office of the *Spring* is with its Point at *g*, to thruft
the *Matrice* clofe againft the *Carriage* and *Body*.

¶ 10. *Of*

¶ 10. *Of the* Hooks, *or* Haggs.

Thefe are *Iron Wyers* about a *Long Primmer* thick: Their Shape you may fee in the Figure: They are fo faftned into the Wood of the *Mold,* that they may not hinder the *Ladle* hitting the *Mouth.* Their Office is to pick and draw with their Points the *Break* and *Letter* out of the *Mold* when they may chance to ftick.

¶ 11. *Of the* Woods *of the* Mold.

All the *Iron* Work aforefaid of the *Mold* is fitted and faftned on two Woods, *viz.* each half one, and each Wood about an Inch thick, and of the fhape of each refpective *Bottom-Plate.* The Wood hath all its Sides except the hind-fide, about a *Pica* longer than the *Bottom-Plate;* but the hind-fide lies even with the *Bottom-Plate.* The *Bottom-Plate,* as afore was faid in ¶ 2. of this §. hath an *Iron Pin* on its under fide, about half an Inch long, with a *Male-fcrew* on its end, which *Pin* being let fit into an Hole in the Wood does by a *Nut* with a *Female-fcrew* in it draw, all the *Iron* Work clofe and faft to the Wood.

But becaufe the Wood is an Inch thick, and the *Pin* in the *Bottom-Plate* but half an Inch long, therefore the outer or under fide of the Wood (as pofited in the Figure) hath a wide round Hole made in it flat at the Bottom, to reach within an *Englifh,* or a *Great Primmer* of the upper fide of the Wood. This round Hole is wide enough to receive the *Nut* with the *Fe-male-*

male-fcrew in it; and the *Pin* being now long enough to receive the *Female-fcrew* at the wide Hole, the *Female-fcrew* may with *round nofed Plyers* be turned about the *Male-fcrew* on the *Pin* aforefaid, till it draw all the *Iron* Work clofe to the Wood.

The Wood behind on the upper half is cut away as the *Bottom-Plate* of that half is; and into the thicknefs of the Wood, clofe by the right and left-hand fide of this *Notch* is a fmall fquare *Wyer-ftaple* driven, which we may call the *Matrice-Check*; for its Office is only to keep the *Shanck* of the *Matrice* from flying out of this *Notch* of the *Mold* when the *Cafter* is at Work. And the *Nuts* and *Screws* of the *Carriage* and *Mouth-piece*, &c. that lie under the *Bottom-Plate*, are with fmall *Chiffels* let into the upper fide of the Wood, that the *Bottom-plates* may lie flat on it.

Sect. XVI.   *Of juftifying the* Mold.

ALthough the *Mold* be now made; nay, very well and Workman-like made, yet is it not imagin'd to be fit to go to work withal; as well becaufe it will doubtlefs Rag (as *Founders* call it; for which Explanation fee the Table) as becaufe the Body, Thicknefs, Straightnefs, and length of the *Shanck* muft be finifht with fuch great Nicety, that without feveral Proofs and Tryings, it cannot be expected to be perfectly true.

Therefore before the finking and juftifying the *Matrices*, the *Mold* muft firft be *Juftified*: And firft, he juftifies the *Body*, which to do, he cafts about twenty *Proofs* or Letters, as they are called, though it matters

matters not whether the *Shancks* have yet Letters
on them or no. Thefe *Proofs* he fets up in a *Compofing-ftick*, as is defcribed in § 17. ¶ 2. Plate 19. at G,
with all their *Nicks* towards the right Hand, and
then fets up fo many Letters of the fame *Body*, (which
for Diftinction-fake we will call *Patterns*) that he
will juftifie his *Body* too, upon the *Proofs*, with all
their *Nicks* alfo to the right Hand, to try if they a-
gree in length with the fame Number of Letters that
he ufes for his *Pattern*; which if they do not, for very feldom they do, but by the Workman's fore-caft
are generally fomewhat too big in the *Body*, that
there may be Subftance left to *Juftifie* the *Mold*, and
clear it from *Ragging*. Therefore the *Proofs* may
drive out fomewhat, either half a Line (which in
Founders and *Printers* Language is half a *Body*) or a
whole Line. (more or lefs.)

He alfo tries if the two fides of the *Body* are parallel, *viz.* That the *Body* be no bigger at the *Head* than
at the *Foot*; and that he tries by taking half the number of his *Proofs*, and turning the Heads of them
lays them upon the other half of his *Proofs*, fo that
if then the *Heads* and *Feet* be exactly even upon
each other, and that the Heads and Feet neither *drive
out*, nor *get in*, (*Founders* and *Printers* Language, for
which fee the Table) the two fides of the *Body* are
parallel; but if either the *Head* or *Foot drives
out*, the two fides of the *Body* are not parallel, and
muft therefore be mended.

And as he has examin'd the Sides of the *Body* fo alfo he examines the thicknefs of the Letter, and tries
if the two Sides of the thicknefs be alfo parallel, which

to

12

to do, he fets up his *Prooves* in the *Compofing-ftick* with their *Nicks* upwards. Then taking half of the *Prooves*, he turns the *Heads* and lay the *Heads* upon the *Feet* of the other half of his *Prooves*, and if the *Heads* and *Feet* lies exactly upon each other and neither *drive-out* or *get-in* the two Sides of the thickneffes are parallel. But if either the *Head* or *Foot* *drive-out* the two Sides of the thickneffes are not parallel; and muft therefore be mended.

Next, he confiders whether the fides of the *Body* be ftraight, firft by laying two Letters with their *Nicks* upwards upon one another, and holding them up in his Fingers, between his Eye and the Light, tries if he can fee Light between them: For if the leaft Light appear between them, the *Carriage* is not ftraight. Then he lays the *Nicks* againft one another, and holds them alfo againft the Light, as before: Then he lays both the *Nicks* outward, and examines them that way, that he may find whether either or both of the *Carriages* are out of ftraight.

But we will fuppofe now the *Body* fomewhat too big, and that it drives out at the *Head* or *Foot*; and that the thicknefs *drives-out* at the *Head* or *Foot* and that the Sides of the *Body* are not ftraight. Thefe are Faults enough to take the *Mold* afunder: but yet if there were but one of thefe Faults it muft be taken afunder for that; by unfcrewing the *Male-Gage*, to take the *Body* off the *Carriage*, and the *Carriage* off the *Bottom-Plate*.

Having found where the Fault of one or both fides of the *Body* is, he lays the *Body* down upon the *Ufing File*; and if the Fault be extuberant, he rubs the

Extu-

Extuberancy down, by preffing his Finger or Fingers hard upon the oppofite fide of the Body, juft over the extuberant part; and fo rubbing the Body hard forwards on the *Ufing-File*, and drawing it lightly backwards, he rubs till he has wrought down the extuberancy, which he examins by applying the *Lyner* to that fide of the Body, and holding it fo up between his Eye and the Light, tries whether or not the *Lyner* ride upon the part that was extuberant; which if it do, the extuberancy is not fufficiently rub'd off, and the former Procefs muft again begin and be continued till the extuberancy be rub'd off. And if the Body were too big, he by this Operation works it down: Becaufe the extuberancy of the *Body* rid upon the *Carriage*, and bore it up.

And if the fault be a *Dawk*, or Hollow in the *Body*, then he Works the reft of that fide of the *Body* down to the bottom of the *Dawk*, which by applying the *Lyner* (as afore) he tryes, and this alfo leffens the *Body*.

If the *Body* drive-out at *Head* or *Foot*, he lays the weight of his Fingers heavy at that fide or end of the *Body* which is too thick, and fo rubs that down harder.

If the thicknefs of the Letter, drive-out at *Head*, or *Foot*, he Screws the *Body* into the *Vice*, and with a flat fharp *File*, files the *Side* down at the *Head*, or *Foot*. At the fame time, if the *Shanck* of the Letter be not Square, he mends that alfo, and fmooth-files it very well.

Then

Then he puts the *Mold* together again: And melting, (or laying afide) his firft *Proofs*, left they fhould make him miftake, he again *Cafts* about twenty New *Proofs*, and examins by them as before, how well he has mended the *Body*, and how near he has brought the *Body* to the fize of the *Pattern*: For he does not expect to do it the *Firft*, *Second*, or *Seventh* time; but mends on, on, on, by a little at a time, till at laft it is fo finifht.

If the *Body* prove too fmall, it is underlaid with a thick or a thin *Affidue*; or fometimes a thin *Plate* of *Brafs*.

Then he examins the *Mouth-piece*, and fees that the *Jaws* flide exactly true, upon every part of the *Pallat* without riding.

If the *Throat* of the *Mouth-piece* lie too low, as moft commonly it is defigned fo to do; Then a *Plate* of *Brafs* of a proper thicknefs is laid under it to raife it higher.

He alfo Juftifies the *Regifters*, making their *Cheeks* truly Square. And Screwing them about an n from the Corner of the *Body*.

He tryes that the *Male* and *Female-Gages* fit each other exactly, and lie directly ftraight along, and parallel to both the Sides of the *Carriage*.

All this thus performed he needs not (perhaps) take the *Mold* affunder again. But not having yet confider'd, or examin'd the length of the *Shanck* of the Letter, he now does; and if it be fomwhat too long (as we will fuppofe by forecaft it is) then the *Body* and *Carriage* being Screwed together, and both the Halves fitted in their *Gages*, the Edges of the
Carri-

Carriage and *Body* are thus together rub'd upon the
Uſing-File, till the *Carriage* be brought to an exact
length.

Having thus (as he hopes) finiſht the juſtifying
of the *Mold*; and put it together, and Screwed it faſt
up, he puts the two Halves together, and then
Rubs or Slides them hard againſt one another, to
try if he can perceive any little part of the *Body* Ride
upon the *Carriage*, or *Carriage* ride upon rhe *Body*:
To know which of them it is that Rides, or is extuberant, he uſes the *Liner*; applying it to both the Places, as well of the *Body* as the *Carriage*: where he ſees
they have Rub'd or bore upon one another: And
which of them that is extuberant, the Edge of the *Liner*
will ſhew, by Riding upon it: And that part he Files
upon with a ſmall flat and very fine *File*, by little
and little, taking off the extuberancy, till the *Bodies*
and *Carriages* lie exactly flat upon, and cloſe to one
another: Which if they do not, the *Mold* will be ſure
to *Rag*.

§. XVII. ¶ 1. *Of Sinking the* Punches *into the*
Matrices.

THat the *Matrice*, and all its parts may be the
better underſtood, as I ſhall have Occaſion
to Name them, I have given you a *Draft* of the *Matrice* in Plate, 18 at E. and ſhall here explain its parts.

E The *Matrice*, wherein is Punched E, the *Face* of
the *Letter*.

a The *Bottom* of the *Matrice*.

b The *Top* of the *Matrice*.

c The

12*

c The *Right Side* of the *Matrice.*
d The *Left Side* of the *Matrice.*
f g The *Face* of the *Matrice.*
h i The *Leather Grove* of the *Matrice.*

In the *Back* or *Side* behind the *Matrice,* juft be-hind E is filed in athwart the *Back,* from the right to the left Side a *Notch,* to fettle and hold the point of the *Spring* or *Wyer* of the *Mold* in, that the *Matrice* fly or ftart not back when it is at Work.

As I told you (in §. 11. ¶ 1.) that the *Punches* are to be made of feveral Thickneffes, for reafons there fhewed; and that therefore the *Letter-Cutter* makes *Wooden Patterns* for his feveral Sizes of Thickneffes as well as Heights; fo now I am come to the *Sinking* of the *Punches* into the *Matrices,* I muft tell you again that the *Letter-Cutter* or elfe the *Founder,* (either of which that *Sinks* them; for fometimes it is a Task Incumbent on each of them) confiders the Thickneffes of all the *Punches* he has to *Sink,* though Heighth he need not confider in *Sinking* the *Matrices*: For the *Matrices,* by reafon of their length in *Copper* upwards and downwards, have Subftance enough and to fpare, for the longeft Letters to be *Sunk* into them: Therefore I fay, he only confiders the feveral Thickneffes of all the *Punches,* and makes *Wooden Patterns* for them, marking with a Pen and Inck the number of each fize, on the *Pattern* as before he did for the *Steel-Punches*: But the *Patterns* he made for the *Steel Punches* will be too Thin for the *Copper Matrices*: Becaufe the *Steel Punches* by *Sinking* into the *Matrices,* ftretch and force the Sides of the *Copper* out, and fometimes crack them for want of Subftance

ftance; and at other times carry or force the Subftance of the *Matrice* fo low with their *Sholder* if the
Letter be broad, that it creates a great Trouble to rub
them *Flat*, (as it is called) becaufe it is done upon
the *Ufing-File*.

Therefore he makes *Wooden Patterns* for every
of the former fiz'd *Punches*, fo thick or rather an
n thicker at the leaft, then he made the *Wooden*
Patterns, that the *Steel-Punches* were made to be
Forged by, that there may be Subftance enough
on each-fide the *Copper* to bear the dilating that
the finking of the *Punch* into it will make, becaufe the *Counter-Puncht-Letters* are Thicker by their
Stems and *Footing* or *Topping* than the *Counter-Punches* made for them need be.

Therefore (as before) for three fizes of *Punches* to be *Counter-Puncht*, he made three feveral
fiz'd *Patterns*; fo now for the feveral fiz'd *Punches* that are to be *Sunk* into *Matrices*, he makes
three feveral fiz'd *Patterns* of *Wood* for the *Copper-
Smith* to draw out *Rods* of *Copper* of thofe feveral
Sizes by, and each of them (as aforefaid) an n, and
for the Thick *Letters* an m (at leaft) Thicker than
the *Patterns* were made, for the *Steel-Punches* to
be Forged to a fize by.

In the Forging of thefe *Copper Rods*, he in-
ftructs the *Copper-Smith* to make Choice of the
fofteft *Copper* he can get, that the *Steel-Punches*
may run the lefs hazzard of breaking ; and
fometimes (if too foft Temper'd) battering their
Stroaks.

The *Rofe Copper* is commonly accounted the fofteft :
But

But yet I have many times *Sunk Punches* indifferently into every fort of *Copper.* Nay, even caſt *Copper*, which is generally accounted the Hardeſt: Becauſe *Copper*, as well (as ſome other Mettals) Hardens with Melting.

Theſe *Rods* of *Copper* are (as I told you in §. III. ¶ 1. to be Cut into ſmall Lengths, each about an Inch and an half long, and a *Great-Primmer* or *Double-Pica* deep; and for great Bodyed *Letters* a *Two-lin'd-Engliſh* deep; But their Thickneſs not aſſignable, becauſe of the Different Thickneſſes in *Letters*, both of the ſame and other *Bodies*, as in part I ſhewed, in §. II. and more fully in this preſent §. and ¶.

The reaſon why the *Copper-Rods* are Forg'd ſo deep, is, That the more ſubſtance of *Copper* may lie under the *Face* of the *Punch*: For if the *Rod* have not a convenient depth, the *Face* of the *Punch* in *Sinking*, does the ſooner ingage with the Hardneſs of the *Face* of the *Stake* it is *Sunk* upon: And having with a few Blows of the *Hammer*, ſoon hardned the *Copper* juſt under the *Face* of the *Punch*, as well the hardneſs of the ſmall (thus hardned) *Body* of *Copper* juſt under the *Face* of the *Punch*, as the Hardneſs of the *Face* of the *Stake* contribute a complycated aſſiſtance to the breaking or battering the *Face* of the *Punch.* But if the *Rod* be deep, the Subſtance of *Copper* between the *Face* of the *Punch* and the *Stake* is leſs hardned, and conſequently the *Punch* will *Sink* the eaſier, and deeper with leſs Violence.

But

But fometimes it has happ'ned that for the *Sinking* one *Matrice* or two, I have been loath to trouble my felf to go to the *Copper-Smiths*, to get one Forg'd: and therefore I have made fhift with fuch *Copper* as I have had by me. But when it has not been fo deep as I could have wifht it, I have juft entred the *Punch* into the *Matrice* upon the *Stake*, and to *Sink* it deep enough, I have laid it upon a good thick piece of *Lead*, which by reafon of its foftnefs has not hardned the *Copper* juft under the *Face* of the *Punch*; but fuffered the *Punch* to do its *Office* with good Succefs.

Having cut the *Copper-Rods*, into fit Lengths with a Cold *Chiffel*, He files the end that is to ftand upon the *Stool* of the *Mold* exactly fquare, and the Right-fide of the *Matrice*, that ftands againft the *Carriage* and *Body*, alfo exactly Square and fmooth upon the *Ufing-File*. Then he places the filed end, or *Bottom* upon the *Stool*, with the *Face* of the *Matrice* towards the *Carriage* and *Body*, and the Right fide of the *Matrice*, clofe againft the *Regifter*: Then if the *Punch* to be funck be an afcending Letter. He with a fine pointed *Needle*, makes a fmall Race by the upper fide of the *Carriage* upon the *Face* of the *Matrice*, and that Race is a mark for him, to fet the top of the Afcending *Letter* at, when he *Sinks* it into the *Matrice*: So that then placing the *Punch* upright upon the middle of the Thicknefs of the *Matrice*, the *Matrice* lying folid on the *Stake*: He with the *Face* of an *Hammer* fizable to the bignefs of his *Punch*, cautioufly knocks upon the *Hammer-end* of the *Punch*, with reiterated Blows, till he he have driven the *Punch* deep enough into the *Matrice.*

But if it be a fhort *Letter,* or a Defcending *Letter,* and not Afcending alfo: Then he elects any *Caft-Letter* of the Thicknefs of the *Beard,* (as *Founders* and *Printers* call it) For which Explanation fee the Table, and he lays that *Letter* upon the *Surface* of the *Carriage,* and then placing the *Bottom* of the *Matrice* to be *Sunk* as before, on the *Stool,* and againft the *Regifter,* He draws with a *Needle* as before, a race above the *Surface* of that *Letter,* againft the *Face* of the *Matrice,* and that race is a Mark for him to place the *Head* of the *Letter* by. Then managing the *Punch* and *Hammer* as before was fhewed, he *Sinks* the *Punch* into the *Matrice.*

But here arifes a Queftion, *viz.* How deep the *Punches* are to be *Sunk* into the *Matrices?* The Anfwer is, a Thick *Space* deep, though deeper even to an n would be yet better: Becaufe the deeper the *Punches* are *Sunk,* the lower does the *Beards* ftand below the *Face,* and thofe *Beards* when the *Caft Letter* comes into the *Printers* Hands to be ufed, are the lefs fubject to *Print,* as too oft they do both at *Head* or *Foot* of a *Page,* than when they lie fo high that the foftnefs of the *Blankets,* and Hardnefs of a *Pull,* or elfe carelefnefs of Running the *Carriage* of the *Prefs* to a confidered Mark they would be. But they are feldom *Sunk* any deeper then a thick Space: and the reafon is, becaufe the breaking or battering the *Face* of the *Punch* fhould not be to much hazarded.

The

The many *Punches* to be *Sunk* into *Matrices* for
the fame *Body*, are difficult to be *Sunk* of an equal
depth. Therefore I always make a *Beard-Gage*, as is
defcribed in *Plate* 19 at F, where *a b* is a *Sholder*
that refts upon the *Face* of the *Matrice*, *c* is the
Point or *Gage* that meafures the depth of the *Sunken Punch*. So that when the *Point c* juft touches the *Bottom*, and both the *Sholders a b* the *Face*
of the *Matrice*, the *Punch* may be accounted well
Sunk as to depth.

But though it be accounted well *Sunk* for a
firft Effay, yet can it not be reafonably imagined
it is well *Sunk* for good and all; as well becaufe in
Sinking the *Punches* it has carryed fome part of
the *Surface* of the *Matrice* down below the *Face* of
the *Matrice* into the *Body* of the *Copper*, as becaufe
both the *Sides* are doubtlefs extorted, and one Side
or Part of the *Punch Sunk* more or lefs deeper than
the other. Wherefore I now come to

¶ 2. *Juftifying the* Matrices.

Juftifying of *Matrices* is, 1. to make the *Face* of
the *Sunken* Letter, lie an exact defigned depth below the *Face* of the *Matrice*, and on all its fides
equally deep from the *Face* of the *Matrice*. 2. It is
to fet or *Juftifie* the *Foot-line* of the Letter exactly
in *Line*. 3. It is to *Juftifie* both the fides, *viz*.
the Right and left-fides of a *Matrice* to an exact
thicknefs.

Therefore to proceed Methodically, he firft flightly Files down the *Bunchings* out that the *Punch*
made

made in the Sides of the *Matrice*; And then flightly
Files down all the *Copper*, on the *Face* of the *Matrice*, till the Hollow the *Punch* made becomes e-
ven with the whole *Face* of the *Matrice*.

Then he *Cafts* a *Proof-Letter* or two, and *Rubs*
them: And with the Edge of a Knife cuts out
what may remain in the bottom of the *Shanck*
by reafon of the un-even breaking, off of the
Break that the fquare bottom of the *Shanck* may
not be born off the *Bottom-Ledge* of the *Lining-
Stick*.

But having till now faid nothing of the *Lining-
Stick*, it is proper before I proceed, to give a De-
fcription of it: It is delineated in *Plate* 19 at G.
Where G is the *Plain*, *a* the *Side-Ledge*, *b* the *Bottom-
Ledge*, *c* the *Stilt*, all made of *Brafs*.

The *Plain* is exactly Flat, Straight, and Smooth, that
the *Shancks* of the *Letter* being likewife fo, may lie
flat and folidly on it. Its depth between the *Bot-
tom-Ledg*, and the fore edge is about the length of the
Shanck of the Letter: But the whole *Plain* of *Brafs*
is yet deeper; Becaufe the *Bottom-Ledge* is faftned on it. The *Lining-Stick* is about two *Inches*
long for fmall Letters; but longer for Big-*Bodyed
Letters*.

Both *Bottom* and *Side-Ledge*, is a thin piece of *Brafs*,
from a *Scaboard* to a *Pica* thick, according as the *Body*
whofe *Face* and *Foot-line* is to be *Juftified* in it is
bigger or lefs. Thefe two *Ledges* is an Infide Square
exactly wrought, and with fmall *Rivets* fafted on
the *Side* edge, and on the *Bottom* edge.

The

The *Stilt* is a thin flat piece of *Brafs-Plate* a-bout a *Scaboard* thick, and a *Double-Pica* broad: One of its edges is *Soldered* to the under-fide of the *Plain*, about a *Double-Pica* within the fore-edge of the *Plain*, that the *Lining-Stick* (when fet by with *Proof-Letters* in it) may not lie flat on its *Bottom*; but have its fore edge *Tilted* up, that the *Letters* in it may reft againft the *Bottom-Ledge*.

Having cut the *Notch* in the *Break* of the *Let-ters* as aforefaid, He *Rubs* every fide of them on the *Stone*, with two or three hard *Rubs*, to take off the fmall *Rags* that may happen on the *Shanck* of the *Letter*, notwithftanding the *Mold* is imagined to be very truly made and *Juftified*.

The *Stone* is commonly a whole *Grind-Stone*, about eighteen Inches diameter, having both its fides tru-ly *Rub'd* flat and fmooth, by *Joftling* it (as *Mafons* call it) upon another broad long and flat Stone with *Sand* and *Water*. It muft have a fine, but very fharp *Greet*. Now to return.

He places a Quadrat of the fame *Body*, on the *Plain* of the *Lining-ftick*, and againft the *Side-Ledge* of it He fets up three or four old m's of the fame *Body*: Then fets up his *Proof-Letter* or *Letters*, and after his *Proof-Letter* three or four old m's more of the fame *Body*; and being very careful that the *Foot* of the *Shanck* of the Letter ftands full down againft the *Bottom-Ledge* of the *Lining-ftick*, He applies the edge of the *Liner* to the *Faces* of all thefe Letters: And if he finds that the edge of the *Liner* juft touch (and no more) as well all the parts of his *Proof-Let-*

ters

ters as they do upon his old *Letters*, He concludes his *Matrice* is *Sunk* to a true *Height againſt
Paper*.

But he feldom hopes for fo good luck; but does
more likely expect the *Matrice* is *Sunk* too deep or
too fhallow, and awry on the right and left-fide,
or on the top or bottom of the *Line*, for all or any
of thefe Faults the *Liner* will eafily difcover. Therefore I fhall fhew you how he *Juſtifies* a *Matrice* that is
too *High againſt Paper*.

We will fuppofe the *Face* of the *Punch* is *Sunk*
flat and ftraight down into the *Matrice*; but yet it
is a little too deep *Sunk*. Therefore he confiders
how much it is too deep: If it be but a little too deep,
perhaps when the *Face* of the *Matrice* fhall be made
exactly flat (for yet it is but *Rough-Filed*) it may
be wrought down to be juft of an *Height againſt Paper*. But if the *Punch* be *Sunk* fo much too deep
that the fmoothing the flat of the *Face* on the *Uſing-
File* will not work it low enough; then with a
Baſtard-cut flat-File, he takes off (according to his
Difcretion) fo much *Copper* from the *Face* of the
Matrice as will make it fo much nearer as he thinks
it wants to the *Face* of the Letter. But yet confiders
that the *Face* of the *Matrice* is yet to fmoothen on
the *Uſing-File*, and therefore he is careful not to
take too muck off the *Face* of the *Matrice* with the
Rough-File.

He is alfo very careful that when he is to *File* upon the *Face* of the *Matrice*, to *Screw* the *Face* of it
Horizontally flat in the *Vice*: And that in *Filing*
upon it, he keeps his *File* directly Horizontal, as
was

was fhewed, *Numb.* 1. *Fol.* 15, 16. *Vol.* 1.   For if he let his right or left-Hand dip, the *File* will in its Natural Progrefs take too much off the fide it dips upon, and confequently the *Face* of the Letter on that fide will lie fhallower from the *Face* of the *Matrice* then it will on the oppofite fide.   The like caution he makes, in *Filing* between the *Top* and *Bottom* of the *Matrice* on the *Face.*   For if he *Files* away too much *Copper* toward the *Top* or *Bottom*, the *Face* of the *Letter* on its *Top* or *Bottom-Line*, will lie on that end fhallower from the *Face* of the *Matrice.*

Then he confiders by his *Proof-Letters* how much too thick the right or left fide of the *Matrice* is.

I told you in § 11. ¶ 4. that the Angle the *Sholder* made with the *Face* of the *Letter*, is about 100 Degrees, which is 10 Degrees more then a *right Angle* or *Square.*   So that if a *Letter* be *Caft* and *Rub'd* juft fo thick that the *Liner* when applied to the *Shanck* of the *Letter* reaches juft to the *Sholder*, there will be an *Angle* of 10 Degrees, contained between the edge of the *Liner* and the *Straight Line* that proceeds from the *Sholder* at the *Shanck*, to the outer-edge of the *Face* of the *Letter.*   And if two *Letters* be thus *Caft* and *Rub'd* and *Set* together, the *Angle* contained between their *Shancks*, and the outer-edge of the *Face* of the *Letter* will be 20 Degrees, which is too wide by half for the *Faces* of two *Letters* to ftand affunder.   Therefore the fides of the *Matrice* muft be fo *Juftified*, that when the *Shancks* of two *Letters* ftand clofe together, the *Angle* between

tween both the *Shancks*, and the adjacent outer-edges of the *Faces* of the *Letters* may both make an *Angle* of about 10 Degrees as aforefaid, which is a convenient diftance for two *Letters* to ftand affunder at the *Face*. But to do which, If the right-fide be too thick, the *Regifter* of the under-half of the *Mold*, being (as I faid) hard fcrew'd, fo as to ftand about an n off the edge of the *Body* towards the right hand; He places the *Foot* of the *Matrice* on the *Stool*, and the right-fide of the *Matrice* clofe againft the *Regifter*, and obferves how much too thick that fide of the *Matrice* is: For fo much as the right-hand edge of the *Orifice* of the *Matrice* ftands on the left hand fide of the *Body*, fo much is the right fide of the *Matrice* too thick, and muft by feveral offers be *Filed* away with a *Baftard-Cut-File*, not all at once, leaft (ere he be aware) he makes that fide of the *Matrice* too thin, which will be a great dammage to the *Matrice*, and cannot be mended but with a *Botch*, as fhall in proper place be fhewed.

Having by feveral proffers wrought the right-fide of the *Matrice* thus near its thicknefs, he proceeds to *Juftifie* the left-fide alfo. But this fide muft be *Juftified* by the upper half of the *Mold*; By turning the top of the *Matrice* downwards, and placing the left-fide of it (now the right-fide) againft the *Regifter*, and works away the left-fide in all refpects as he did the right-fide; ftill being very cautious he takes not to much *Copper* away at once.

To *Juftifie* the *Letter* in *Line* he examins the *Proof-Letter* (yet ftanding in the *Lining-Stick*) and applies
the

the *Liner* to the *Foot-line*: And if the *Liner* touch
all the way upon the *Foot-line* of the *Proof-Letter* and the *Foot-Line* of all the old m's, that *Matrice* is *Juſtified* in *Line*. But this alſo very rarely
happens at firſt, for by deſign it is generally made
to ſtand too low in *Line*: Becauſe the *Bottom* of
the *Matrice* may by ſeveral proffers be *Filed* away
till the *Letter* ſtand exactly in *Line*. But ſhould
he take too much off the *Bottom* of the *Matrice*,
it cannot be made to ſtand lower without another
Botch.

Nor does he reckon that this firſt Operation, or
perhaps ſeveral more ſuch, ſhall *Juſtifie* the *Matrice* in *Line*. But after bringing both the ſides
of the *Matrice* thus near, and alſo bringing the
Matrice thus near the *Line*. He *Caſts* another *Proof-
Letter* or two, and *Rubbing* all the ſides of their
Shancks, as before was ſhew'd, he tries by *Rubbing*
the *Letters* how near he has brought the thickneſs
of both the ſides: For when the ſides of the *Matrice* are brought juſt to ſuch a thickneſs, that the
Shanck of the *Letter* (*Caſt* in the *Mold*) *Rubs* flat half
way up beyond the *Beard* towards the *Face* of the
Letter, the *Matrice* is of a convenient thickneſs, and
there the *Angle* from the *Beard* of the *Shanck*, to the
outer-edge of two *Letters* ſet together, will make an
Angle of about 10 degrees as aforeſaid, which being
about one third part of a *thin-Space* is a convenient
diſtance for the adjacent edges of two *Letters* to
ſtand aſſunder: But yet *Founders* ſometimes to *Get in*
or *Drive out*, *Caſt* the *Letters* thinner or thicker, and
conſequently their *Faces* ſtand cloſer or wider aſſunder

13

der, which is unfeemly when the *Letter* comes to be *Printed*.

Then he fets the *Proof-Letters* in the *Lining-Stick*, between four or five old m's as before, and with the *Liner* examins again how well thefe *Proof-Letters* ftand in *Line* with the old m's, which if they do not, he Reiterates the former Operations fo oft, till the fides and *Line* of the *Matrice* is *Juftified*, and at every Operation *Cafts* new *Proof-Letters* to examine the thicknefs of both the Sides, and how well the *Matrice* is *Juftified* to *Stand in Line*.

The *Matrice* being now *Juftified*, he *Files* a *Leather-Groove* round about it, *viz* a *Notch* (made propereft with a three fquare *File*) within about a thick *Scaboard* of the top of the *Matrice*, to tie the *Leather* faft to.

He alfo *Files* another *Notch* in the back-fide of the *Matrice* athwart it, to reft the point of the *Wyer* or *Spring* in. But this *Notch* muft by no means be made before the *Matrice* be *Juftified* to its true *Height againft Paper*: Becaufe when this *Notch* is made, the *Punch* cannot again be ftruck in the *Matrice*; For that the *Matrice* will not lie folid on the *Stake* in that place.

¶ 3. *Of* Botching-Matrices, *to make them ferve the better.*

Matrices are fometimes either through a carelefs, or fometimes through an unlucky ftroak or two of the *File* made too thin. And fometimes the *Foot* of the *Matrice* is too much taken away, and the Letter by
that

that means ftands too high in *Line*: And fometimes the *Face* of the *Matrices* is too much taken away; So that the Letter will not ftand *High enough againft Paper*.

To remedy all or any part of thefe inconveniencies, *Founders* are forced to make *Botches* on the *Matrice*: As firft, If the *Matrice* be too thin on the right or left fide, or both; They prick up that fide, by laying the *Matrice* flat on the *Work-Bench*, with the thin fide upwards, and holding the point of a *Punch-Graver* aflope upon the thin fide, with an *Hammer* drive the point into the thin fide of the *Matrice*, and fo raife a *Bur* upon that fide; which *Bur* (though it thicken not the *Matrice*, yet it) makes the fide of the *Matrice* ftand off the *Regifter*, and confequently is equevalent to thickning it.

The higher this *Bur* is raifed, the better is the *Matrice Botcht*; becaufe the thin fine points thus raifed (if not pretty well flatted into the Subftance of the *Bur*) will quickly either wear off by the preffure of the *Regifter* againft them, or elfe flatten into the *Body* of the *Bur*, and both ways makes the *Matrice* again too thin.

Sometimes they do not *Botch* the *Matrice* thus for this fault; but only Pafte a piece of Paper, or a Card, (according as it may want thicknefs) againft the thin fide of the *Matrice* and fo thicken it.

But to mend the fides I ufe another Expedient, *viz.* by Soldering a piece of *Plate-Brafs* againft its thin fide or fides, which is much better than *Botching* it.

Second-

Secondly, If the *Matrice* be filed away too much at the *Foot,* they knock it up with the *Pen* of the *Hammer*; and ftretch it between the *Foot* and the *Orifice* of the *Matrice,* and then *Juftifie* it again in *Line.* Or a piece may be *Soldered* under the *Foot.*

Thirdly, If the *Face* of the *Matrice* be too much taken away, and either the *Punch* fpoiled or the *Notch* in the back of the *Matrice* made fo, as it cannot be *Sunken* deeper, they raife a *Bur* on the *Face,* as they did on the thin fides, to keep the *Matrice* off the *Carriages* and *Bodies* which Lengthens the height of the Letter *againft Paper* fo much as is the height of the raifed *Bur.* But of all the *Botches* this is the worft, becaufe the *Beard* lies now nearer the *Face*: And the hollow ftanding off of the *Face* of the *Matrice* from the *Carriages* and *Bodies,* fubjects the Mettal to run between them, and fo pefters the Workman to get the Letter out of the *Mold* and *Matrice.*

Sect. XVIII, *Of fetting up the* Furnance.

HAving *Juftified* the *Mold* and *Matrice,* we come now to *Cafting* of *Letters*: But yet we have neither *Furnance, Mettal,* or *Ladle.* Wherefore it is the *Founders* care, firft to provide thefe.

The *Furnance* I have defcribed in Plate 20. It is built of Brick upright, with four fquare fides and a Stone on the top, in which Stone is a wide round hole for the *Pan* to ftand in.

a b c d The

a b c d The fquare Stone at the top, covering the whole *Furnance*. This is indeed the *Furnance*.

a d, b c The breadth two Foot and one Inch.

a b, c d The Length two Foot three Inches. Into the Breadth and Length about the whole Stone, is let in even with the top of the Stone a fquare *Iron Band* two Inches deep, and a quarter and half quarter of an Inch thick to preferve the Edges of the Stone from battering.

e The round hole the *Pan* ftands in, which hath an *Iron Plate* let into it eight Inches diameter, an Inch and half broad and one quarter of an Inch thick.

This *Iron*-Plate fits the *infide* of the *Hole* fo far as it is Circular, and confequently is a *Segment* of a *Circle*. But where the *Smoak-vent* breaks off the Circularity of the Stone, there ends this Plate of *Iron*, that the Smoak may have the freer vent. Its Office alfo is to preferve the Edge of the *Hole* from battering, with the oft taking out and putting in the *Iron Pan*.

f The *Funnel* feven Inches high, and five Inches wide.

g The *Stoke-Hole* four Inches wide, and fix Inches long.

h h The height of the *Furnance* two Foot ten Inches.

i The *Air-Hole* juft underneath the Hearth to let in Air that the Fire may burn the freer.

k The *Afh-Hole* where the Afhes that fall from the Hearth are taken away.

l m n o The

13*

l m n o The *Bench* two Foot broad, three Foot long, and two Foot eight Inches high. The *Bench* is to empty the Letters out of the *Mold* upon, as the *Founder Casts* them.

The *Hearth* lies seven Inches below the top of the round *Hole*, and hath under it another round *Iron-Ring* of the same demensions with the first, on which straight *Iron-Bars* are fastened that the *Fire* is laid on.

In the round *Iron-Ring* (or rather Segment) on the top of the *Furnance* is set the *Pan*, which is either a *Plate Ladle*, or a small *Cast-Iron Kettle* that sinks into it within two Inches of the *Brims* of the *Pan*.

¶ 2. *Of making* Mettal.

The Mettal *Founders* make *Printing Letters* of, is *Lead* hardned with *Iron*: Thus they chuse *stub-Nails* for the best *Iron* to Melt, as well becaufe they are a-fured *stub-Nails* are made of good soft and tough *Iron*, as becaufe (they being in small pieces of *Iron*) will Melt the sooner.

To make the *Iron Run*, they mingle an equal weight of *Antimony* (beaten in an Iron-Morter in-to small pieces) and *stub-Nails* together. And pre-paring so many Earthen forty or fifty pounds *Melt-ing-Pots* (made for that purpofe to endure the *Fire*) as they intend to ufe: They *Charge* thefe Pots with the mingled *Iron* and *Antimony* as full as they will hold.

Every

Every time they Melt *Mettal,* they build a new *Furnance* to melt it in: This *Furnance* is called an *Open Furnance*; becaufe the Air blows in through all its fides to Fan the *Fire*: They make it of Bricks in a broad open place, as well becaufe the Air may have free accefs to all its fides, as that the Vapours of the *Antimony* (which are Obnoxious) may the lefs offend thofe that officiate at the *Making* the *Mettal*: And alfo becaufe the Violent Fire made in the *Furnance* fhould not endanger the Firing any adjacent Houfes.

They confider before they make the *Furnance* how many Pots of *Mettal* they intend to Melt, and make the *Furnance* fizable to that number: We will fuppofe *five Pots.* Therefore they firft make a Circle on the Ground capable to hold thefe five *Pots,* and wider yet by three or four Inches round about: Then within this Circle they lay a Courfe of Bricks clofe to one another to fill the Plain of that Platform, with their broad or flat fides downwards, and their ends all one way, and on this Courfe of Bricks they lay another Courfe of Bricks as before, only the Lengths of this Courfe of Bricks lies athwart the Breadths of the other Courfe of Bricks: Then they lay a third Courfe of Bricks with their lengths crofs the Breadth of the fecond Courfe of Bricks.

Having thus raifed a Platform, they place thefe five *Pots* in the middle of it clofe to one another, and then on the Foundation or Plat-form raife the *Furnance* round about by laying the Bricks of the firft *Lay* end to end and flat, clofe to one another:

On

On the fecond *Lay*, they place the middle of a Brick
over a *Joynt* (as *Brick-layers* call it) that is where
the ends of two Bricks joyn together, and fo again
lay Bricks end to end till they *Trim* round the *Plat-
form*. Then they lay a third *Lay* of Bricks, covering
the *Joynts* of the fecond *Lay* of Bricks as before: So
is the Foundation finifht.

Then they raife the Walls to the *Furnance* on this
Foundation; But do not lay the ends of their Bricks
clofe together. But lay the ends of each Brick about
three Inches off each other, to ferve for *Wind-holes*
till they *Trim* round about: Then they lay another
Lay of Briks leaving other fuch *Wind-holes* over the
middle of the laft *Lay* of Bricks, and fo *Trim* as
they work round either with half Bricks or Bats that
the *Wind-holes* of the laft *Lay* may be covered:
And in this manner and order they lay fo many *Lays*
till the Walls of the *Furnance* be raifed about three
Bricks higher than the *Mouths* of the *Melting-Pots*,
ftill obferving to leave fuch *Wind-holes* over the
middle of every Brick that lies under each *Lay*.

Then they fill the fides of the *Furnance* round a-
bout the *Melting-Pots*, and over them with *Char-coal*,
and *Fire* it at feveral *Wind-holes* in the bottom till
it burn up and all over the *Furnance*, which a mo-
derate Wind in about an Hours time will do: And
about half an Hours time after they lay their Ears
near the Ground and liften to hear a *Bubling* in the
Pots; and this they do fo often till they do hear it.
When they hear this *Bubling*, they conclude the *Iron*
is melted: But yet they will let it ftand, perhaps
half an hour longer or more, according as they guefs
the

the Fire to be Hotter or Cooler, that they may be the more aſſured it is all throughly Melted. And when it is Melted the Melting *Pot* will not be a quarter full.

And in or againſt that time they make another ſmall *Furnance* cloſe to the firſt, (to ſet an *Iron Pot* in, in which they Melt *Lead*) on that ſide from whence the Wind blows; Becauſe the Perſon that Lades the *Lead* out of the *Iron-Pot* (as ſhall be ſhewed by and by) may be the leſs annoyed with the Fumes of the *Mettal*, in both *Furnances*. This *Furnance* is made of three or four *Courſe* of Bricks open to the windward, and wide enough to contain the deſigned *Iron Pot*, with room between it and the ſides to hold a convenient quantity of *Charcoal* under it, and about it.

Into this *Iron-Pot* they put for every three Pound of *Iron*, about five and twenty pounds of *Lead*. And ſetting Fire to the *Coals* in this little *Furnance* they Melt and Heat this *Lead* Red-hot.

Hitherto a Man (nay, a Boy) might officiate all this Work; But now comes Labour would make *Hercules* ſweat. Now they fall to pulling down ſo much of the ſide of the open *Furnance* as ſtands above the Mouth of that *Melting-Pot* next the *Iron-Pot*, And having a thick ſtrong *Iron Ladle*, whoſe *Handle* is about two Yards long, and the *Ladle* big enough to hold about ten Pounds of *Lead*, and this *Ladle* Red-hot that it chill not the *Mettal*, they now I ſay with this *Ladle* fall to clearing this firſt *Melting-Pot* of all the Coals or filth that lie on the top of the Melted *Mettal*: while another Man at the ſame time
ſtand

ftands provided with a long ftrong round *Iron Stir-ring Poot*; the *Handle* of which *Stirring Poot* is al-fo about two Yards long or more, and the *Poot* it felf almoft twice the length of the depth of the Mel-ting *Pot*. This *Poot* is nothing but a piece of the fame *Iron* turned to a fquare with the Handle: And this *Poot* is alfo in a readinefs heated Red-hot.

Now one Man with the *Ladle Lades* the *Lead* out of the *Iron-Pot* into the Melting *Pot*, while the o-ther Man with the *Poot* ftirs and Labours the *Lead* and *Mettal* in the Melting *Pot* together till they think the *Lead* and *Mettal* in the Melting *Pot* be well incorporated: And thus they continue *Lading* and *Stirring* till they have near filled the Melting *Pot*.

Then they go to another next *Melting-Pot*, and fucceffively to all, and Lade and ftir *Lead* into them as they did into the firft. Which done the *Mettal* is made: And they pull down the *Walls* of the *Open Furnance*, and rake away the Fire that the *Mettal* may cool in the *Pots*.

Now (according to Cuftom) is Half a Pint of Sack mingled with Sallad Oyl, provided for each Work-man to Drink; intended for an Antidote againft the Poyfonous Fumes of the *Antimony*, and to re-ftore the Spirits that fo Violent a Fire and Hard Labour may have exhaufted.

¶ 3. *Of*

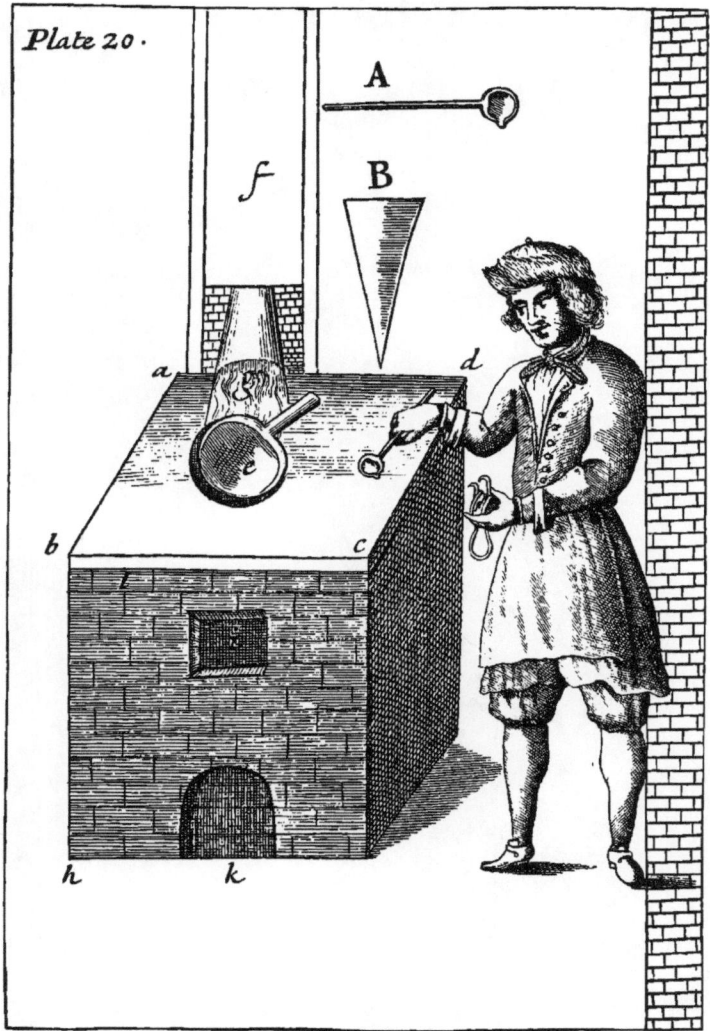

Plate 20.

¶ 3. *Of* Letter-Ladles.

Letter-Ladles differ nothing from other common *Ladles*, fave in the fize: Yet I have given you a Draft of one in Plate 20 at A. Of thefe the *Cafter* has many at Hand, and many of feveral fizes that he may fucceffively chufe one to fit the feveral fizes of *Letters* he has to *Caft*; as well in *Bodies* as in *Thickneffes*.

§ XIX. ¶ 1. *Of* Cafting, Breaking, Rubbing, Kerning, *and fetting up of* Letters.

BEfore the *Cafter* begins to *Caft* he muft kindle his *Fire* in the *Furnance*, to *Melt* the *Mettal* in the *Pan*. Therefore he takes the *Pan* out of the Hole in the Stone, and there lays in *Coals* and kindles them. And when it is well kindled, he fets the *Pan* in again, and puts *Mettal* into it to *Melt*. If it be a fmall *Bodyed-Letter* he *Cafts*, or a thin *Letter* of Great *Bodies*, his *Mettal* muft be very hot; nay, fometimes Red-hot to make the Letter *Come*. Then having chofe a *Ladle* that will hold about fo much as the *Letter* and *Break* is, he lays it at the *Stokinghole*, where the Flame burfts out to heat. Then he ties a thin Leather cut into fuch a Figure as is defcribed in Plate 20 at B with its narrow end againft the *Face* to the *Leather-Groove* of the *Matrice*, by whipping a Brown Thred twice about the *Leather-Groove*, and faftning the Thred with a Knot. Then he puts both Halves of the *Mold* together, and puts
                                                    the

the *Matrice* into the *Matrice Cheek*, and places the *Foot* of the *Matrice* on the *Stool* of the *Mold*, and the broad end of the *Leather* upon the *Wood* of the upper half of the *Mold*, but not tight up, left it might hinder the *Foot* of the *Matrice* from *Sinking* clofe down upon the *Stool* in a train of Work. Then laying a little Rofin on the upper *Wood* of the *Mold*, and having his *Cafting Ladle* hot, he with the bolling fide of it Melts the Rofin; And when it is yet *Melted* preffes the broad end of the *Leather* hard down on the *Wood*, and fo faftens it to the *Wood*. All this is Preparation.

Now he comes to *Cafting*. Wherefore placing the under-half of the *Mold* in his left hand, with the *Hook* or *Hag* forward, he clutches the ends of its *Wood* between the lower part of the *Ball* of his Thumb and his three hind-Fingers. Then he lays the upper half the *Mold* upon the under half, fo as the *Male-Gages* may fall into the *Female-Gages*, and at the fame time the *Foot* of the *Matrice* place it felf upon the *Stool*. And clafping his left-hand Thumb ftrong over the upper half of the *Mold*, he nimbly catches hold of the *Bow* or *Spring* with his right-hand Fingers at the top of it, and his Thumb under it, and places the point of it againft the middle of the *Notch* in the backfide of the *Matrice*, preffing it as well forwards towards the *Mold*, as downwards by the *Sholder* of the *Notch* clofe upon the *Stool*, while at the fame time with his hinder-Fingers as a-forefaid, he draws the under-half of the *Mold* towards the *Ball* of his Thumb, and thrufts by the *Ball* of his Thumb the upper part towards his Fingers,

gers, that both the *Regiſters* of the *Mold* may preſs againſt both ſides of the *Matrice,* and his Thumb and Fingers preſs both Halves of the *Mold* cloſe together.

Then he takes the Handle of his *Ladle* in his right Hand, and with the *Boll* of it gives a ſtroak two or three outwards upon the *Surface* of the *Melted Mettal* to ſcum or cleer it from the Film or Duſt that may ſwim upon it. Then takes up the *Ladle* full of *Mettal,* and having his *Mold* as aforeſaid in his left hand, he a little twiſts the left-ſide of his *Body* from the *Furnance,* and brings the *Geat* of his *Ladle* (full of *Mettal*) to the *Mouth* of the *Mold,* and twiſts the upper part of his right-hand towards him to turn the *Mettal* into it, while at the ſame moment of Time he Jilts the *Mold* in his left hand forwards to receive the *Mettal* with a ſtrong *Shake* (as it is call'd) not only into the *Bodies* of the *Mold,* but while the *Mettal* is yet hot, running ſwift and ſtrongly into the very *Face* of the *Matrice* to receive its perfect Form there, as well as in the *Shanck.*

Then he takes the upper half of the *Mold* off the under half, by placing his right-Hand Thumb on the end of the *Wood* next his left-Hand Thumb, and his two middle Fingers at the other end of the *Wood,* and finding the Letter and *Break* lie in the under-Half of the *Mold* (as moſt commonly by reaſon of its weight it does) he throws or toſſes the Letter *Break* and all upon a Sheet of Waſte Paper laid for that purpoſe on the *Bench* juſt a little beyond his left-hand, and is then ready to *Caſt* another Letter as before,

fore, and alſo the whole number that is to be *Caſt*
with that *Matrice*.

But ſometimes it happens that by a *Shake*, or too
big a *Ladle*, the Mettal may ſpill or ſlabber over the
Mouth of the upper Half of the *Mold*, ſo that the
ſpilt *Mettal* ſticking about the out-ſides of the *Mouth*,
may lift the Letter off the under half of the *Mold*, and
keep it in the upper half.   Therefore he with the
point of the *Hag* in the Wood of the under half of
the *Mold*, picks at the hollow in the fore part of the
Break made by the *Shaking* out of the *Mettal*, and
draws *Break* and *Letter* both out.   It ſometimes ſticks
in the under Half of the *Mold* by the ſame cauſe,
and then he uſes the point of the *Hag* in the
upper half of the *Mold*, to pick or hale it out, as
before.

It alſo ſometimes ſticks when any of the Joynts
of the *Mold* open never ſo little, the *Mettal* thus get-
ting in between thoſe Joynts: But this fault is not to
be indured, for before he can *Caſt* any more, this fault
muſt be mended.

But beſides *Letters*, there is to be *Caſt* for a perfeƈt *Fount* (properly a Fund) *Spaces* Thick and Thin,
n *Quadrats*, m *Quadrats* and *Quadrats*.   Theſe are
not *Caſt* with *Matrices* but with *Stops* (as we may
call them) Becauſe when theſe are *Caſt* they are all
ſhorter than the *Shanck* of the Letter, that they may
not *Print*.   Therefore they take off the *Regiſter* of
the under-Half *Mold*, and fit a piece of *Plate-Braſs*
about a *Brevier* Thick and a *Brevier* longer than to
reach to the edge of the *Body* in the place of the *Regiſter*, and drill a hole in this *Plate-Braſs* right againſt
                                        the

the Hole in the *Carriage* that the *Female-Screw* lies
in: This Hole is made ſo wide that the *Male-Screw*
which ſcrewed the *Regiſter* cloſe to the *Carriage* and
Body may enter in at it, and ſcrew this *Plate-Braſs*
cloſe to them, as it did the *Regiſter*: Then they make
a mark with the point of a *Needle* on the *Plate-Braſs*
juſt againſt the ſide of the Edge of the *Body*, and at
this mark they double down the end of the *Plate-
Braſs* inwards to make a perfeƈt *Square* with the *inſide* of the whole *Plate*. This doubling down is called the *Stop* aforeſaid, and muſt be made juſt ſo thick
as they deſign the Thin or Thick *Space* to be, and
muſt have its Upper and Under-Edges filed ſo exaƈtly to the *Body*, that it may lie cloſe upon the
Under-*Carriage*, and juſt even ſo high as the upper-
ſide of the *Body*. So that when the Upper-half of
the *Mold* is placed on the under-Half, and *Mettal
Caſt* in at the *Mouth* (as before) the *Mettal* ſhall de-
ſcend no deeper between the two *Bodies* then juſt
to his *Stop*: You muſt note that this *Stop* muſt be
filed exaƈtly true as to *Body* and *Thickneſs*: For if
it be never ſo little too big in *Body*, the *Carriage* of
the *Mold* will ride upon it and make the *Body* of
the *Space* bigger. Or if the *Body* be never ſo little
too little, the Hot *Mettal* will run beyond the *Stop*;
both which Miſcarriages in making the *Stop*, ſpoil the
Space.

  If the *Space* be too ſhort, they File the end of the
Stop ſhorter.

  This *Brevier* thick *Plate* will be thick enough for
Stops for the Thin or Thick *Spaces* of any Body
though of *Great-Cannon*, and for the n *Quadrat Stop*
                                                    of

14

of any Body under a *Great Primmer*.   And for the
m *Quadrat Stop* of all to a *Brevier* and all Bodies under it.   But for *Stops* that require to be Thicker
then a *Brevier*, inftead of doubling the *Stop* inwards
on the *Plate*, I *Solder* on the in-fide of that end of
the *Plate* a *Stop* full big enough in Body, and big
enough in Thickneſs for the *Quadrat* I intend to
make, and afterwards file and fit the *Stop* exaċtly as
before.

When they *Caſt* thefe *Spaces* or *Quadrats*, this *Stop*
is always fcrewed faſt upon the *Carriage* of the under-Half *Mold* as aforefaid.   So that they only fit
the upper half *Mold* on the under, and *Caſt* their
Number almoſt twice as quick as they do the Letters
in *Matrices*.

It is generally obferved by *Work-men* as a Rule,
That when they *Caſt Quadrats* they *Caſt* them exaċt-
ly to the Thickneſs of a fet Number of m's or *Body*,
viz. two m's thick, three m's thick, four m's thick,
&c.   And therefore the *Stops* aforefaid muſt all be
filed exaċtly to their feveral intended thickneſſes,
The reafon is, that when the *Compofiter* Indents any
Number of Lines, he may have *Quadrats* fo exaċtly
Caſt that he fhall not need to *Juſtifie* them either with
Spaces or other helps.

¶ 2. *Some Rules and Circumſtances to be obferved in*
Caſting.

1. If the Letter be a fmall *Body*, it requires a
Harder *Shake* than a great *Body* does: Or if it be
a thin Letter though of a greater *Body*, efpecially
fmall

fmall *i*, being a thin Letter its Tittle will hardly *Come*;
So that fometimes the *Cafter* is forced to put a little
Block-Tin into his Mettal, which makes the Mettal
Thinner, and confequently have a freer flux to the
Face of the *Matrice*.

2. He often examines the *Regifters* of the *Mold*,
by often *Rubbing* a *Caft* Letter: For notwithftanding the *Regifters* were carefully *Juftified* before,
and hard fcrewed up; yet the conftant thrufting
of both *Regifters* againft the fides of the *Matrice*,
may and often do force them more or lefs to drive
backwards.   Or a fall of one half or both Halfs of
the *Mold*, may drive them backwards or forwards:
Therefore he examins, as I faid, how they *Rub*, whether
too Thick or too Thin.   And if he fee Caufe, mends
the *Regifters*, as I fhew'd § 5. ¶ 2.

Or if the *Matrice* be *Botcht*, as I fhew'd you § 5.
¶ 3. then thofe *Botches* (being only fo many fine
points rifing out of the Body of the *Copper* of the
Matrice) may with fo many reiterated preffures of
the *Regifters* againft them, flatten more and more,
and prefs towards the Body of the *Matrice*, and confequently make the Letter Thinner: Which if it do,
this muft be mended in the *Matrice* by re-raifing it to
its due Thicknefs.

3. He pretty often examins, as I fhew'd in § 5.
¶ 2. how the Letters ftand in *Line*: For when great
Numbers are *Caft* with one *Matrice*, partly by preffing the point of the *Wyer* againft the *Bottom-Sholder*
of the *Notch* in the back-fide of the *Matrice*, and
partly by the foftnefs of the matter of his *Matrice* and
hardnefs of the *Iron-ftool*, the *Foot* of the *Matrice* (if
it

it wear not) may batter ſo much as to put the Letter out of Line. This muſt be mended with a *Botch*, *viz.* by knocking up the *Foot* of the *Matrice*, as I ſhew'd § 5.·¶ 3.

A Work-man will *Caſt* about four thouſand of theſe Letters ordinarily in one day.

<h3 style="text-align:center">¶ 3. *Of* Breaking *off* Letters.</h3>

Breaking off is commonly Boys-work: It is only to *Break* the *Break* from the *Shanck* of the *Letter*. All the care in it is, that he take up the *Letter* by its Thickneſs, not its *Body* (unleſs its Thickneſs be e-qual to its Body) with the fore Finger and Thumb of his right Hand as cloſe to the *Break* as he can, leſt if when the *Break* be between the fore-Finger and Thumb of his left Hand, the force of *Breaking* off the *Break* ſhould bow the *Shanck* of the *Let-ter*.

<h3 style="text-align:center">¶ 4. *Of* Rubbing *of* Letters.</h3>

Rubbing of *Letters* is alſo moſt commonly Boys-work: But when they do it, they provide *Finger-ſtalls* for the two fore-Fingers of the right-Hand: For elſe the Skin of their Fingers would quickly rub off with the ſharp greet of the Stone. Theſe *Fin-ger-ſtalls* are made of old *Ball-Leather* or *Pelts* that *Printers* have done with: Then having an heap of one ſort of *Letters* lying upon the Stone before them, with the left-Hand they pick up the *Letter* to be *Rub'd*, and lay it down in the *Rubbing* place with
one

one of its fides upwards they clap the Balls of the fore-Finger and middle-Finger upon the fore and hinder-ends of the *Letter*, and *Rubbing* the *Letter* pretty lightly backwards about eight or nine Inches, they bring it forwards again with an hard preffing *Rub* upon the *Stone*; where the fore-Finger and Thumb of the left-Hand is ready to receive it, and quickly turn the oppofite fide of the *Letter*, to take fuch a *Rub* as the other fide had.

But in *Rubbing* they are very careful that they prefs the Balls of their Fingers equally hard on the *Head* and *Foot* of the *Letter*. For if the *Head* and *Foot* be not equally preft on the *Stone*, either the *Head* or *Foot* will *Drive out* when the *Letters* come to be *Compofed* in the *Stick*; So that without *Rubbing* over again they cannot be *Dreft*.

¶ 5. *Of* Kerning *of* Letters.

Amongft the *Italick-Letters* many are to be *Kern'd*, fome only on one fide, and fome both fides. The *Kern'd-Letters* are fuch as have part of their *Face* hanging over one fide or both fides of their *Shanck*: Thefe cannot be *Rub'd*, becaufe part of the *Face* would *Rub* away when the whole fide of the *Shanck* is toucht by the *Stone*: Therefore they muft be *Kern'd*, as *Founders* call it: Which to do, they provide a fmall Stick bigger or lefs, according as the *Body* of the *Letter* that is to be *Kern'd*. This *Kerning-ftick* is fomewhat more than an Handful long, and it matters not whether it be fquare or round: But if it be fquare the Edges of it muft be pretty

14*

ty well rounded away, left with long usage and hard
Cutting they Gall the Hand. The upper side of this
Kerning-Stick is flatted away somewhat more than
the length of the *Letter*, and on that flat part is
cut away a flat bottom with two square sides like
the *Sides* or *Ledges* of the *Lining-stick* to serve for
two *Sholders*. That side to be *Kern'd* and *scrap'd*, is
laid upwards, and its opposite side on the bottom of
the *Kerning-stick* with the *Foot* of the *Letter* against
the bottom *Sholder*, and the side of the *Letter* against
the side *Sholder* of the *Kerning-stick*.

He also provides a *Kerning-Knife*: This is a pretty strong piece of a broken Knife, about three Inches long, which he fits into a Wooden-Handle: But
first he breaks off the Back of the Knife towards the
Point, so as the whole edge lying in a straight-line
the piece broken off from the back to the edge may
leave an angle at the point of about 45 Degrees,
which irregular breaking (for so we must suppose
it) he either *Grinds* or *Rubs* off on a *Grind-stone*.
Then he takes a piece of a Broom-stick for his Handle, and splits one end of it about two Inches long
towards the other end, and the split part he either
Cuts or Rasps away about a *Brevier* deep round a-
bout that end of the Handle. Then he puts about
an Inch and an half of his broken blade into the
split or slit in the Handle, and ties a four or five
doubled Paper a little below the Rasped part of
the Handle round about it, to either a *Pica* or
Long-Primmer thick of the slit end of the Handle. This *Paper* is so ordered that all its sides
round about shall stand equally distant from all
the

the Rafped part of the Handle: For then fetting
the other end of the Handle in Clay, or otherwife faftening it upright, when *Mettal* is poured
in between the Rafped part of the Handle and
the Paper about it, that *Mettal* will make a ftrong
Ferril to the *Handle* of the *Knife*. The irregularities that may happen in *Cafting* this *Ferril*
may be Rafped away to make it more handy and
Handfome.

Now to return again where I left off. Holding
the Handle of the *Kerning-ftick* in his left-Hand,
He lays the fide of the *Letter* to be *Kern'd* upwards with the *Face* of the *Letter* towards the
end of the *Kerning-ftick*: the fide of the *Letter*
againft the fide *Sholder* of the *Kerning-ftick*, and
the *Foot* of the *Letter* againft the bottom *Sholder*
of the *Kerning-ftick*, and laying the end of the
Ball of his left-Hand Thumb hard upon the *Shanck*
of the *Letter* to keep its *Side* and *Foot* fteddy
againft the *Sholders* of the *Kerning-ftick*, he with
the *Kerning-Knife* in his right-Hand cuts off about
one quarter of the *Mettal* between the *Beard* of
the *Shanck* and the *Face* of the *Letter*. Then turning his *Knife* fo as the back of it may lean towards him, he fcrapes towards him with the edge of
the *Knife* about half the length of that upper-fide,
viz. about fo much as his Thumb does not cover:
Then he turns the *Face* of the *Letter* againft the
lower *Sholder* of the *Kerning-ftick*, and fcraping
fromwards him with a ftroak or two of his *Knife*
fmoothens that end of the *Letter* alfo.

If the other fide of the *Letter* be not to be *Kern'd*
it

it was before *Rub'd* on the *Stone*, as was fhewed
in the laft ¶ : But if it be to be *Kern'd*, then he
makes a little hole in his *Kerning-ftick*, clofe to the
lower *Sholder* of it and full deep enough to receive
all that part of the *Face* of the *Letter* that hangs o-
ver the *Shanck*, that the *Shanck* of the *Letter* may lie
flat and folid on the bottom of the *Kerning-ftick*,
and that fo the *Shanck* of the *Letter* bow not when
the weight of the Hand preffes the edge of the
Kerning-Knife hard upon it.   Into this hole he puts
(as before faid) fo much of the *Face* of the *Letter*
as hangs over the fide of the *Shanck*, and fo
fcrapes the lower end of the *Letter* and *Kerns*
the upper end, as he did the former fide of the
Letter.

¶ 6. *Of* Setting up, *or* Compofing Letters.

I defcribed in § 5. ¶ 2. the *Lining-ftick*, But
now we are come to *Setting up*, or *Compofing* of
Letters.   The *Founder* muft provide many *Compofing-fticks*; five or fix dozen at the leaft.   Thefe
Compofing-fticks are indeed but long *Lining-fticks*,
about feven or eight and twenty Inches long *Handle* and all: Whereof the *Handle* is about three
Inches and an half long: But as the *Lining-ftick*
I defcribed was made of *Brafs*: So thefe *Compofing-fticks*
are made of *Beech-Wood*.

When the Boy *Sets* up *Letters* (for it is commonly Boys Work) The *Cafter Cafts* about an hundred *Quadrats* of the fame *Body* about half an Inch
broad at leaft, let the *Body* be what it will, and of
the

the length of the whole *Carriage*, only by placing
a flat *Brafs* or *Iron Plate* upon the *Stool* of the *Mold*
clofe againft the *Carriage* and *Body*, to ftop the *Mettal*
from running farther.

The Boy (I fay) takes the *Compofing-ftick* by
the *Handle* in his left-Hand, clafping it about
with his four Fingers, and puts the *Quadrat* firft
into the *Compofing-ftick*, and lays the Ball of his
Thumb upon it, and with the fore-Finger and Thumb
of his right-Hand, affifted by his middle Finger to
turn the *Letter* to a proper pofition, with its *Nick* upwards towards the bottom fide of the *Compofing-ftick*;
while it is coming to the *Stick*, he at the fame time
lifts up the Thumb of his left-Hand, and with it receives and holds the *Letter* againft the fore-fide of
the *Quadrat*, and after it, all the *Letters* of the fame
fort, if the *Stick* will hold them, If not he *Sets* them
in fo many *Sticks* as will hold them: Obferving to
Set all the *Nicks* of them upwards, as aforefaid.
And as he *Set* a *Quadrat* at the beginning of the
Compofing-ftick, fo he fils not his *Stick* fo full, but
that he may *Set* another fuch *Quadrat* at the end
of it.

¶ 7. *Some Rules and Circumftances to be obferved in*
Setting *up* Letters.

1. If they *Drive* a little out at *Head* or *Foot*, fo
little as not to require new *Rubbing* again, then
he holds his Thumb harder againft the *Head* or *Foot*,
fo as to draw the *Driving* end inward: For elfe when
they come to *Scraping*, and *Dreffing* the *Hook* of the
Dref-

Dreffing-Hook drawing Square, will endanger the
middle or fome other part of *Letters* in the *Stick*
to *Spring* out: And when they come into the *Dreffing-block*, the *Knots* of the *Blocks* drawing alfo
fquare fubject them to the fame inconvenience. And
if they *Drive* out at the *Head*, the *Feet* will more or
lefs ftand off one another: So that when the *Tooth*
of the *Plow* comes to *Drefs* the *Feet*, it will more or
lefs job againft every *Letter*, and be apt to make a
bowing at the *Feet*, or at leaft make a *Bur* on their
fides at the *Feet*.

2. When *Short-Letters* are begun to be *Set* up in a
Stick, the whole *Stick* muft be fill'd with *Short-
Letters*: Becaufe when they are *Dreffing*, the *Short
Letters* muft be *Bearded* on both fides the *Body*:
And fhould *Short-Letters* or *Afcending* or *Defcending*
or *Long* ftand together, the *Short* cannot be *Bearded*
becaufe the *Stems* of the *Afcending* or *Defcending*
or *Long-Letters* reach upon the *Body* to the *Beard*:
So that the *Short-Letters* cannot be *Bearded*, unlefs
the *Stems* of the other *Letters* fhould be fcra-
ped off.

3. When *Long-Letters* are begun to be *Set* up in the
Stick, none but fuch muft fill it, for the reafon a-
forefaid.

4. If any *Letters Kern'd* on one fide be to be
Set up, and the *Stems* of the fame *Letters* reach
not to the oppofite *Beard* as f or f, in *Setting* up
thefe or fuch like *Letters*, every next *Letter* is turned with its *Nick* downwards, that the *Kern* of
each *Letter* may lie over the *Beard* of its next. But
then they muft be all *Set* up again with a *Short-
Letter*

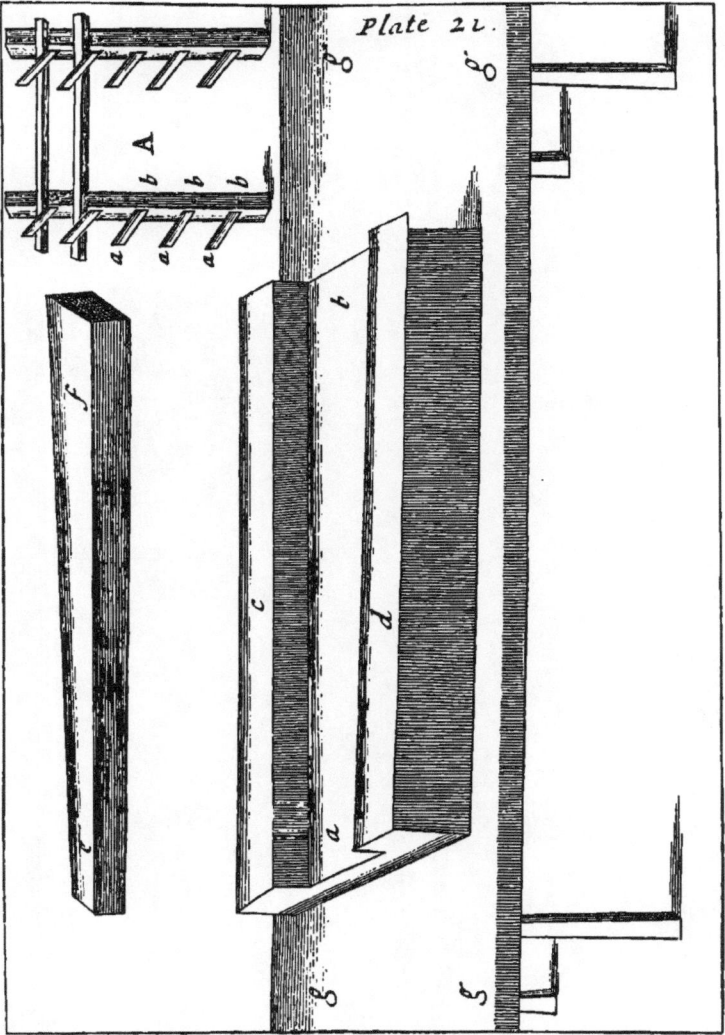

Plate 22.

Letter between each, that they may be *Bearded*.

As every *Stick-full* is ſet up, he ſets them by upon the *Racks*, ready for the *Dreſſer* to *Dreſs*, as ſhall be ſhewed in the next §.

The *Racks* are deſcribed in *Plate* 21. at A. They are made of Square *Deal Battens* about ſeven Inches and an half long, as at *a b a b a b*, and are at the ends *b b b* let into two upright *Stiles*, ſtanding about ſixteen Inches and an half aſſunder, and the fore-ends of the *Racks* mounting a little, that when *Sticks* of *Letters* is *Set* by on any two parallel *Racks*, there may be no danger that the *Letters* in them ſhall ſlide off forward; but their *Feet* reſt againſt the *Bottom-Ledges* of the *Compoſing-ſticks*. They ſet by as many of theſe *Sticks* with *Letter* in them, as will ſtand upon one another between every two *Rails*, and then ſet another pile of *Sticks* with *Letter* in them before the firſt, till the length of the *Rail* be alſo filled with *Sticks* of *Letter* before one another. They ſet all the *Sticks* of *Letters* with their ends even to one another with the *Faces* of the *Letter* forwards.

This *Frame of Racks* is always placed near the *Dreſſing-Bench*, that it may ſtand convenient to the *Letter-Dreſſers* Hand.

§ 20. ¶ 1. *Of* Dreſſing *of* Letters.

THere be ſeveral Tools and Machines uſed to the *Dreſſing* of *Letters*: And unleſs I ſhould deſcribe them to you firſt, you might perhaps in my following diſcourſe not well underſtand me:

me: Wherefore I ſhall begin with them: They
are as follows.

1. The *Dreſſing-Sticks*.
2. The *Bench*, *Blocks* and its Appurtenances.
3. The *Dreſſing-Hook*.
4. The *Dreſſing-Knife*.
5. The *Plow*.
6. The *Mallet*.

Of each of theſe in order.

¶ 2. *Of the* Dreſſing-Sticks.

I need give no other Deſcription of the *Dreſſing-
ſticks*, than I did in the laſt § and ¶ of the *Compoſing-
Sticks*: Only they are made of hard Wood, and of
greater Subſtance, as well becauſe hard Wood will
work ſmoother than ſoft Wood, as becauſe greater
Subſtance is leſs Subject to warp or ſhake than ſmaller Subſtance is.   And alſo becauſe hard Wood is
leſs Subject to be penetrated by the ſharpneſs of
the *Bur* of the *Mettal* on the *Letters* than the
ſoft.

¶ 3. *Of the* Block-Grove, *and its* Appurtenances.

The *Block-Grove* is deſcribed in *Plate* 21. *a b* The
Groove in which the *Blocks* are laid, two Inches
deep, and ſeven Inches and an half wide at one end,
and ſeven Inches wide at the other end: One of the
Cheeks as *c* is three Inches and an half broad at
one end, and three Inches broad at the other
end, and the other *Cheek* three Inches broad the whole
Length

Plate 22.

a

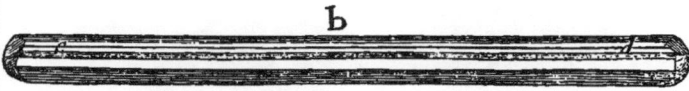

b

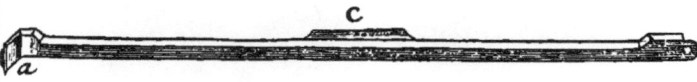

c

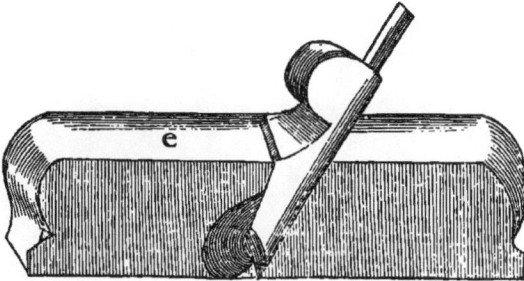

e

d

Length: The Length of theſe *Cheeks* are two and twenty Inches.

The *Wedge e f* is ſeven and twenty Inches and an half long, two Inches broad at one end, and three Inches and an half broad at the other end; And two Inches deep.

g g g g The *Bench* on which the *Dreſſing-Blocks* are placed, are about ſixteen Inches broad, and two Foot ten Inches high from the Floor. The *Bench* hath its farther Side, and both ends, railed about with ſlit Deal about two Inches high, that the *Hook*, the *Knife*, and *Plow*, &c. fall not off when the Work-man is at Work.

The *Blocks* are deſcribed in *Plate* 21 at a b: They are made of hard Wood. Theſe *Blocks* are ſix and twenty Inches long, and each two Inches ſquare. They are *Male* and *Female*, a the *Male*, b the *Fe-male*: Through the whole Length of the *Male-Block* runs a *Tongue* as at *a b*, and a *Groove* as at *c d*, for the *Tongue* of the *Plow* to run in; This *Tongue* is about half an Inch thick, and ſtands out ſquare from the upper and under ſides of the *Block*. About three Inches within the ends of the *Block* is placed a *Knot* as at *c c*: Theſe *Knots* are ſmall ſquare pieces of *Box-wood*, the one above, and the other be-low the *Tongue*.

The *Female Block* is ſuch another *Block* as the *Male Block*, only, inſtead of a *Tongue* running through the length of it a *Groove* is made to receive the *Tongue* of the *Male-Block*, and the *Knots* in this *Block* are made at the contrary ends, that when the *Face* of a *Stick* of *Letter* is placed on the
                                                    Tongue

15

Tongue the *Knot* in the *Male-Block* stops the *Stick* of *Letter* from sliding forwards, while the other *Knot* in the *Female-Block* at the other end, by the knocking of a *Mallet* on the end of the *Block* forces the *Letter* between the *Blocks* forwards, and so the whole *Stick* of *Letters* between these two *Knots* are screwzed together, and by the *Wedge e f* in *Plate* 21 (also with the force of a *Mallet*) *Wedges* the two *Blocks* and the *Stick* of *Letter* in them also tight, and close between the sides of the two *Blocks*; that afterwards the *Plow* may more certainly do its Office upon the *Foot* of the *Letter*; as shall be shewed hereafter.

¶ 3 *Of the* Dreffing-Hook.

The *Dreffing-Hook* is described in *Plate* 21 at c. This is a long square *Rod* of *Iron*, about two Foot long and a *Great-Primmer* square: Its end *a* is about a *two-Lin'd Englifh* thick, and hath a small *Return* piece of *Iron* made square to the under-side of the *Rod*, that when the whole *Dreffing-Hook* is laid along a *Stick* of *Letter*, this *Return piece* or *Hook* may, when the *Rod* is drawn with the *Ball* of the Thumb, by the *Knot* on the upper side of it at c, draw all the *Letter* in the *Stick* tight and close up together, that the *Stick* of *Letter* may be *Scraped*, as shall be shewed.

¶ 4 *Of*

¶ 4 *Of the* Dreſſing-Knife.

The *Dreſſing-Knife* is delineated at d in *Plate.* 21. It is only a ſhort piece of a *Knife* broken off about two Inches from the *Sholder*: But its Edge is *Baſil'd* away from the back to the point pretty ſuddenly to make it the ſtronger: The *Sprig* or *Pin* of the *Handle* is commonly let into an Hole drilled into a piece of the Tip of an Harts-horn, as in the Figure and is faſtned in with *Roſen,* as other *Knives* are into their *Handles.*

¶ 5 *Of the* Plow.

The *Plow* is delineated in *Plate* 21 at e: It is almoſt a common *Plain* (which I have already deſcribed in *Vol.* 1. *Numb.* 4. *Plate* 4. and § 2 to 9.) only with this diſtinction, that through the length of the *Sole* runs ſuch a *Tongue,* as does through the *Male-Block* to ſlide tight and yet eaſily through the *Groove* made on the top of the *Male-block*: Its *Blade* makes an *Angle* of 60 Degrees with the *Sole* of it.

§ 21. ¶ 1. *Of* Dreſſing *of* Letters.

THe *Letter Dreſſer* hath (as I told you before) his *Letter Set* up in *Compoſing-ſticks,* with their *Nicks* upwards, and thoſe *Sticks* ſet upon the *Racks*: Therefore he takes one *Stick* off the *Racks,* and placing the *Handle* of the *Compoſing-ſtick* in his left-hand, he he takes the contrary end of the *Dreſſing-ſtick* in his right-hand, and laying the Back of the *Dreſſing-ſtick* even upon or rather a little hanging over the Back of the *Compoſing-ſtick*, that the *Feet* of the *Letter* may fall within the *Bottom-Ledge* of the *Dreſſing-ſtick*; He at the ſame time fits the *Side-Ledge* of the *Dreſſing-ſtick* againſt the farther end of the *Line* of *Letters* in the *Compoſing-ſtick*: And holding then both *Sticks* together, his left-Hand at the *Handle-*end of the *Compoſing-ſtick*, and his right-Hand within about two Handfuls of the *Handle-*end of the *Dreſſing-ſtick*, He turns his Hands, *Sticks* and all, outward from his left-Hand, till the *Compoſing-ſtick* lies flat upon the *Dreſſing-ſtick*, and conſequently the *Letters* in the *Compoſing-ſtick* is turned and laid upon the *Dreſſing-ſtick*.

Then he goes as near the Light as he can with the *Letters* in his *Dreſſing-ſtick*, and examins what *Letters Come not well* either in the *Face* or *Shanck*: So that then holding the *Dreſſing-ſtick* in his left-Hand, and tilting the *Bottom-Ledge* a little downward, that the *Feet* of the *Letter* may reſt againſt the *Bottom-Ledge*, and laying the Ball of his Thumb upon any certain Number of *Letters* between his *Body* and the *Letter* to be *Caſt out*, He with the *Foot* of a *Space* or ſome thin *Letter*, lifts up the *Letter* to be *Caſt out*, and lets it fall upon the *Dreſſing-Bench*: and thus he does to all the *Letters* in that *Stick* that are to be *Thrown out*.

Then taking again the *Dreſſing-Stick* in his left-Hand at or near the handle of it, he takes the *Dreſſing-Hook* at the *Knot*, between the fore-Finger and
Thumb

Thumb of his right-Hand, and laying the *Hook* over
the edge of the *Quadrat* at the farther end of the
Dreffing-ftick, near the *bottom-Ledge* of it, he flips
his right-Hand to the *Handle* of the *Dreffing-ftick,*
and his left-Hand towards the middle of the *Dreffing-ftick,* fo as the end of the Ball of his Thumb
may draw by the farther end of the *Knot* on the
Dreffing-Hook the whole *Dreffing-Hook,* and the *Hook*
at the end of it the whole *Stick* of *Letter* clofe together towards him; While at the fame time he with
his Fingers clutched about the *Stick* and *Letter,* and
the Thumb-*ball* of his Hand preffes the under flat of
the *Hooking-ftick* clofe againft the *Letter* and *Dreffing-
ftick,* that the *Letter* in the *Stick* may lie faft and manageable.

Then he takes the *Handle* of the *Dreffing-Knife*
in his right-Hand, and inclining the back of it towards his *Body,* that its *Bafil*-edge may *Cut* or *Scrap*
the fmoother, He *Scrapes* twice or thrice upon fo
much of the whole *Line* of *Letters* as lies between
the outer-fide of the *Dreffing-Hook* and the *Face* of
the *Letter.*

But if twice or thrice *Scraping,* have not taken all
the *Bur* or irregularities off fo much of the *Letter* as
he *Scraped* upon, he *Scrapes* yet longer and oftner
till the whole number of *Letters* in the *Dreffing-
ftick* from end to end feems but one intire piece of
Mettal.

Thus is that fide of the fore-part (*viz.* that
part towards the *Face*) of the *Shanck* of the *Body*
finifht.

To *Scrape* the other end of that fide of the *Let-*
ter

15*

ter, viz. that towards the *Feet*; He turns the *Handle* of the *Stick* from him, and removing the *Dreffing-Hook* towards the *Face* of the *Letter* which is already *Scraped,* he places his Thumb againſt the *Knot* of the *Dreffing-Hook,* and preſſes it hard from him, that the *Hook* of the *Dreffing-Hook* being now towards him, may force the whole *Stick* of *Letter* forwards againſt the *Side-Ledge* of the *Dreffing-ſtick*; that ſo the whole *Line* in the *Stick* may lie again the faſter and more manageable: Then he *Scrapes* with the *Dreffing-Knife* as before, till the end of the *Shanck* of the *Letter* towards the *Feet* be alſo *Dreſt.*

Then he lays by his *Dreffing-Hook,* and keeping his *Dreffing-ſtick* of *Letter* ſtill in his left-Hand, he takes a ſecond *Dreffing-ſtick,* with its *Handle* in his right-Hand, and lays the *Side-Ledge* of it againſt the hither ſide of the *Quadrat* at the hither end of the *Dreffing-ſtick,* and the *bottom-Ledge* of the ſecond *Stick* hanging a little over the *Feet* of the *Letter,* that they may be comprehended within the *bottom-Ledge* of the ſecond *Dreffing-ſtick*; and ſo removing his left-Hand towards the middle of both *Dreffing-ſticks,* and claſping them cloſe together, he turns both Hands outwards towards the left, till the *Letter* in the firſt *Dreffing-ſtick* lie upon the ſecond *Dreffing-ſtick,* and then the *Face* of the *Letter* will lie outwards toward the right-Hand, and the *Nicks* upwards. Then he uſes the *Dreffing-Hook* and *Dreffing-Knife* to *Scrape* this ſide the *Line* of *Letter,* as he did before to the other ſide of the *Line* of *Letter*: So ſhall both ſides be *Scraped* and *Dreſt.*

Having thus *Scraped* both the ſides, He takes the
Handle

Handle of the *Dreſſing-ſtick* into his left-Hand, as before, and takes the *Male-block* into his right-Hand,
and placing the *Tongue* of the *Block* againſt the *Face*
of the *Letter* in the *Dreſſing-ſtick*, he alſo places the
Knot of the *Block* againſt the farther ſide of the
Quadrat at the farther end of the *Stick*, and ſo placing his right-Hand underneath the middle of the
Dreſſing-ſtick and *Block*, he turns his Hand outwards towards the left, as before, and transfers the
Letter in the *Dreſſing-ſtick* to the *Male-Block*: Yet he
ſo holds and manages the *Block* that the *Shanck* of
the *Letter* may reſt at once upon the ſide of the *Block*
the *Knot* is placed in, and the *Face* of the *Letter* upon the *Tongue*.

When his *Stick* of *Letters* is thus transfer'd to the
Male-Block, He claps the middle of the *Male-Block*
into his left-Hand, tilting the *Feet* of the *Letter* a little upwards, that the *Face* may reſt upon the *Tongue*,
and then takes about the middle of the *Female-Block*
in his right-Hand, and lays it ſo upon the *Male-Block*,
that the *Tongue* of the *Male-Block* may fall into the
Tongue of the *Female-Block*, and that the *Knot* at
the hither end of the *Female Block* may ſtand againſt
the hither ſide of the *Quadrat* at the hither end of
the *Line* of *Letters*: So that when the *Knot* of the
Male-Block is lightly drawn towards the *Knot* of the
Female-Block, or the *Knot* of the *Female-Block*
lightly thruſt towards the *Knot* of the *Male-Block*,
both *Knots* ſhall ſqueeze the *Letter* cloſe between
them.

Then he graſps both *Blocks* with the *Letter* between them in both his Hands, and lays them in
the

the *Block-Groove*, with the *Feet* of the *Letter* upwards, and the hither fide of the hither *Block* againft the hither *Cheek* of the *Block-Groove*. And putting the *Wedge* into the vacant fpace between the *Blocks* and the further *Cheek* of the *Block-Groove*, he lightly with his right-Hand thrufts up the *Wedge* to force the *Blocks* clofe together, and pinch the *Letter* clofe between the *Blocks*.

Then with the *Balls* of the Fingers of both his Hands, he Patts gently upon the *Feet* of the *Letter*, to prefs all their *Faces* down upon the *Tongue*; which having done, he takes the *Mallet* in his right-Hand, and with it knocks gently upon the head of the *Wedge* to pinch the *Letter* yet clofer to the infides of the *Blocks*. Then he Knocks lightly and fuccef-fively upon the *Knot-ends* of both the *Blocks*, to force the *Letters* yet clofer together. And then again knocks now pretty hard upon the head of the *Wedge*, and alfo pretty hard upon the *Knot-ends* of the *Blocks*, to *Lock* the *Letter* tight and clofe up.

Then he places the *Tongue* of the *Plow* in the upper *Groove* of the *Block*; And having the *Tooth* of the *Iron* fitted in the *Plow*, fo as to fall juft upon the middle of the *Feet* of the *Letter*, he grafps the *Plow* in his right-Hand, placing his Wrift-Ball againft the *Britch* of it, and guiding the fore-end with his left-Hand, flides the *Plow* gently along the whole length of the *Blocks*; fo as the *Tooth* of the *Iron* bears upon the *Feet* of the *Letter*: And if it be a fmall *Letter* he *Plows* upon, the *Tooth* of the *Iron* will have cut a *Groove* deep e-nough through the length of the whole *Block* of *Let-ters*:

ters: But if the *Body* of the *Letter* be great, he re-
itterates his *Traverſes* two three or four times according to the Bigneſs of the *Body* of the *Letter*, till
he have made a *Groove* about a *Space* deep in the *Feet*
of the *Shancks* of the whole *Blocks* of *Letter*, and
have cut off all the irregularities of the *Break*.

Then with a ſmall piece of *Buff* or ſome other ſoft
Leather, he rubs a little upon the *Feet* of the *Letter* to
ſmoothen them.

Then he unlocks the *Blocks* of *Letter*, by knocking
with the *Mallet* upon the ſmall end of the *Wedge*,
and firſt takes the *Wedge* from between the *Blocks* and
Cheeks, and lays it upon the farther *Cheek*, and afterwards takes the *Blocks* with *Letter* in it near both ends
of the *Blocks* between the Fingers and Thumbs of
both his Hands, and turns the hithermoſt *Block* upon
the hithermoſt *Cheek*, and with his Fingers and
Thumbs again lifts off the upper *Block*, leaving the *Letter* on the undermoſt *Block* with its *Face* againſt the
Tongue.

Then taking the *Block* with *Letter* in it in his left-
Hand, he places the *Knot-end* from him, and takes
the *Handle* of the *Dreſſing-ſtick* in his right-Hand, and
lays the *Side-Ledge* of it againſt the hither ſide of the
Quadrat at the hither end, and the *Bottom-ledge* a-
gainſt the *Feet* of the *Letter*, he graſps the *Handle* of
the *Dreſſing-ſtick Block* and all in his left-Hand, and
lays his right-Hand Thumb along the under ſide of
the *Dreſſing-ſtick* about the middle, and with the
Fingers of the ſame Hand graſps the *Block*, and
turning his Hands, *Block*, and *Dreſſing-ſtick* to the
right, transfers the *Letter* in the *Block* upon the *Dreſſing-ſtick*.   Then

Then grafping the *Dreffing-ftick* by the *Handle*
with his left-Hand, he with his right-Hand takes the
Dreffing-Hook by the *Knot*, and lays the infide of
the *Hook* of it againft the farther fide of the *Quadrat*
at the farther end of the *Stick*, and drawing the *Hook*
and *Letter* in the *Dreffing-ftick* with his left Thumb
by the *Knot* clofe up toward him, he refting the
Stick upon the *Dreffing-bench* that he may *Scrape* the
harder upon the *Beard* with the Edge of the *Dreffing-Knife*, *Scrapes* off the *Beard* as near the *Face* as
he dares for fear of fpoiling it, and about a Thick
Space deep at leaft into the *Shanck*.

If the Bottom and Top are both to be *Bearded*, He transfers the *Letter* into another *Dreffing-
ftick*, as hath been fhewed, and *Beards* it alfo as before.

¶ 2. *Some Rules and Circumftances to be obferved in*
Dreffing *of* Letters.

1. The *Letter-Dreffer* ought to be furnifht with
three or four forts of *Dreffing-fticks*, which differ no-
thing from one another fave in the Height of their
Ledges.   The *Ledges* of one pair no higher than a
Scaboard.   This pair of *Sticks* may ferve to *Drefs*,
Pearl, *Nomparel*, and *Brevier*.   Another pair whofe
Ledges may be a *Nomparel* high.   And this pair of
Dreffing-fticks will ferve to *Drefs Brevier*, *Long-Primmer*, and *Pica*: Another pair whofe *Ledges* may be
a *Long-Primmer* high: And thefe *Dreffing-fticks* may
ferve to *Drefs Pica*, *Englifh*, *Great-Primmer*, and
Double-Pica.   And if you will another pair of *Dref-
fing-*

Dreſſing-ſticks, whoſe *Ledges* may be an *Engliſh* High: And theſe *Dreſſing-ſticks* may ſerve to *Dreſs* all big Bodyed *Letters,* even to the Greateſt.

2. As he ought to be furniſht with ſeveral ſorts of *Dreſſing-ſticks* as aforeſaid: So ought he alſo to be furniſht with ſeveral *Blocks,* whoſe *Knots* are to correſpond with the Sizes of the *Ledges* of the *Dreſ-ſing-ſticks,* for the *Dreſſing* of ſeveral *Bodies* as aforeſaid.

3. He ought to be furniſht with three or four *Dreſ-ſing-Hooks,* whoſe *Hooks* ought to be of the ſeveral Depths aforeſaid, to fit and ſuit with the ſeveral *Bodyed-Letters.*

4. He muſt have two *Dreſſing-Knives,* one to lie before the *Blocks* to *Scrape* and *Beard* the *Letter* in the *Sticks,* and the other behind the *Dreſſing-blocks* to uſe when occaſion ſerves to *Scrape* off a ſmall *Bur,* the *Tooth* of the *Plow* may have left upon the *Feet* of the *Letter.* And though one *Dreſſing-Knife* may ſerve to both theſe uſes: Yet when Work-men are in a Train of Work they begrutch the very turning the Body about, or ſtepping one ſtep forward or backward; accounting that it puts them out of their Train, and hinders their riddance of Work.

5. For every *Body* of *Letter* he is to have a particular *Plow,* and the *Tooth* of the *Iron* of each *Plow* is to be made exactly to a ſet bigneſs, the meaſure of which bigneſs is to be taken from the ſize of the *Break* that is to be *Plowed* away. For Example, If it be a *Pearl Body* to be *Plowed,* the breadth of the *Tooth* ought not to be above a thin *Scaboard*: Becauſe the *Break* of that *Body* cannot be bigger, for Reaſons I have

have given before; But the *Tooth* muſt be full broad enough, and rather broader than the *Break,* leſt any of the irregularity of the *Break* ſhould be left upon the *Foot* of the *Letter.*   And ſo for every *Body* he fits the *Tooth* of the *Iron,* full broad enough and a little broader than the ſize of the *Break.* This is one reaſon why for every particular *Body* he ought to have a particular *Plow.*   Another reaſon is.

The *Tooth* of this *Plow* muſt be exactly ſet to a punctual diſtance from the *Tongue* of the *Plow*: For if they ſhould often ſhift *Irons* to the ſeveral *Stocks* of the *Plow,* they would create themſelves by ſhifting more trouble than the price of a *Stock* would compenſate.

A *Fount* of *Letter* being new *Caſt* and *Dreſt,* the Boy *Papers* up each ſort in a *Cartridge* by it ſelf, and puts about an hundred Pounds weight, *viz.* a Porters Burthen into a *Basket* to be ſent to the *Maſter-Printers.*

The *Steel-Punches* being now *Cut,* the *Molds* made, the *Matrices Sunk,* the *Letters Caſt,* and *Dreſt,* the application of theſe *Letters* falls now to the task of the *Compoſiter*; whoſe *Trade* ſhall be (God willing) the Subject of the next *Exerciſes.*

F I N I S.

www.ingramcontent.com/pod-product-compliance
Lightning Source LLC
Chambersburg PA
CBHW021947220326
41599CB00012BA/1336